Advances in Applied Science and Engineering

nova
science publishers
www.novapublishers.com

Advances in Applied Science and Engineering

Research Challenges in Science, Engineering and Technology
M. Shanmuganantham, R. Nagarajan, B.E., M.E., PhD (Author),
S. Kannadhasan, PhD (Author)
2021. ISBN: 978-1-68507-008-3 (Hardcover)
2021. ISBN: 978-1-68507-028-1 (eBook)

COVID-19 Pandemic: Questions, Answers and Hypotheses
A. K. Haghi, PhD
2020. ISBN: 978-1-53618-565-2 (Softcover)
2020. ISBN: 978-1-53618-593-5 (eBook)

Innovations and Challenges in Natural Products Chemistry: Research and Applications
A. K. Haghi, PhD
2020. ISBN: 978-1-53618-650-5 (eBook)

More information about this series can be found at
https://novapublishers.com/product-category/series/advances-in-applied-science-and-engineering/

Bhargav Appasani, Om Prakash Acharya,
Amitkumar Vidyakant Jha
and Nisha Gupta
Editors

Metamaterials for Microwave and Terahertz Applications

Absorbers, Sensors and Filters

nova
science publishers
www.novapublishers.com

https://doi.org/10.52305/APHY8244

NOTICE TO THE READER

Library of Congress Cataloging-in-Publication Data

ISBN: 979-8-88697-330-3

Published by Nova Science Publishers, Inc. † New York

Contents

Preface

The research on advanced materials is moving at a pace to achieve better performance with limited resources and time. Electromagnetic materials made up of materials such as metals and dielectrics, when scaled down to subwavelength dimensions and when arranged in a periodic manner, begin to exhibit some unique properties that cannot be obtained or that cannot be observed using the conventional materials. These materials are known as meta-materials and have attracted the attention of the electromagnetic community because of their myriad applications. They are usually arranged in repeating patterns, at a scales that are smaller than the wavelengths of the phenomena they influence. Their precise shape, geometry, dimensions, orientation and arrangement gives them the ability to influence the electromagnetic waves and achieve benefits that go beyond what is possible with conventional materials. Over the last few years, metamaterials have discovered an unprecedented rise in practical applications in various fields of science and engineering, particularly, in microwave and terahertz frequency regimes.

This book will provide a glimpse about the design of filters, sensors and absorbers in microwave and terahertz frequencies. The main objective of this book is to create awareness among young researchers about this challenging and growing field of metamaterials, and to provide them detailed theory of designing such structures.

Chapter 1

Metamaterials Based Microwave Filters

Anamiya Bhattacharya*
SAC, ISRO, Ahmedabad, Gujarat, India

Abstract

The rapid development of wireless communication systems increases the demand for filtering techniques in broad bandwidth technologies and high-speed data transmission. Therefore, the different filters with their unique merits are being exploited in the field of receiver technology. This chapter discusses the feasibility of different filter structures, including ultra-wideband (UWB) band-pass and band-stop filters. At the same time, design methodology, performance metrics, such as insertion loss, return loss, out-of-band attenuation level, roll-off, etc., and their suitable applications are also broadly investigated. Furthermore, a comparative analysis of different prospect filters is also carried out. Finally, an appendix has been included to furnish some guidelines/steps for designing such filters for different applications.

Keywords: filters, frequency selective surface (FSS) structure, metamaterial (MTM), multimode resonator, split ring resonator (SRR), transmission line (TL) structures

1. Introduction to Metamaterials

Metamaterials are usually artificial periodic structures that demonstrate expedient and not usual electromagnetic (EM) properties by modifying their

---

* Corresponding Author's Email: anamiya2007@gmail.com.

In: Metamaterials for Microwave and Terahertz Applications
Editors: B. Appasani, O. Prakash Acharya, A. Vidyakant Jha et al.
ISBN: 979-8-88697-330-3

permittivity and permeability to control dispersion, reflection and refraction of EM waves. The 'metamaterial' structure has no physical existence under any electromagnetic excitation and the phrase 'meta,' which means beyond, implies the same correctly. Therefore, the term 'metamaterials' ascribes to the materials that showcase unusual electromagnetic properties due to the excitation of electromagnetic waves under certain conditions. In recent years metamaterial (MTM) structures have drawn sole interest among global researchers due to their unusual and exclusive controlled electromagnetic properties. These properties significantly influence the development of microwave devices with reduced size, which further leads to the usage of MTM for various microwave applications. Therefore the wise selection of materials and techniques for designing circuits helps to reduce the dimension and controls the unwanted loss significantly in the direction of wave propagation. The various microwave applications using MTM structures can be found widely in designing microwave devices (such as filters, directional couplers, circulators, etc.), antenna applications, designing absorber structures, and constructing frequency selective surface (FSS) structures, imaging applications, in cloaking and radome applications. In the case of wideband applications, there is still continuous analysis between the Salisbury absorber and Jaumann absorber. This concept is further extended by introducing a multilayer structure or using a thin substrate layer, incorporating lumped elements.

This chapter discusses the filter concept and its subsequent use in the microwave and wireless applications. The immeasurable importance and steady progress in different types of filter technology have introduced a new era in modern communication systems to extract and detect the signal information contained within a specific frequency band. In this context, FSS/MTM-based structures have emerged as a feasible solution in the domain of spatial filters, where filtering property in terms of the passband (BP) or band reject (BS) may be achieved appositely by selecting different geometry orientations and the unit cell dimension. Moreover, the implementation of these EM properties is strongly dependent on available advanced technology which further compromises the best performance and fabrication cost. The combination of all these types of work is further illustrated in subsequent sections in this chapter.

2. Design Approaches for Metamaterial Filters

As it has been mentioned above, in this chapter, different types of metamaterial filters based on resonant-type structures are elaborated, which are made of metamaterial-based transmission lines (TL) concept since their origination in 2003 [1]. The detailed properties of metamaterial structures and their constitutive frequency responses have been discussed and illustrated in [2]. The metamaterial structure/resonator is an additional element in designing a conventional band-pass filter, band-reject filter, low-pass filter and high-pass filter. By properly tuning the resonator length, width and gap between two resonators, elimination of spuriousness and improvement of out-of-band rejection is possible [3-4] to a great extent. At the same time, the performance of filters can be retained without any increment in dimension. In this context, the contribution of Split Ring Resonator (SRR) based structures have been greatly appreciated as it helps to reject the unwanted signal where base elements do not help to generate a controllable transmission band. Therefore, different kinds of unit cells and metamaterial-based resonators are adopted in designing versatile as well as various filters. Several filter designs using different possible metamaterial transmission lines are further explored in subsequent sections.

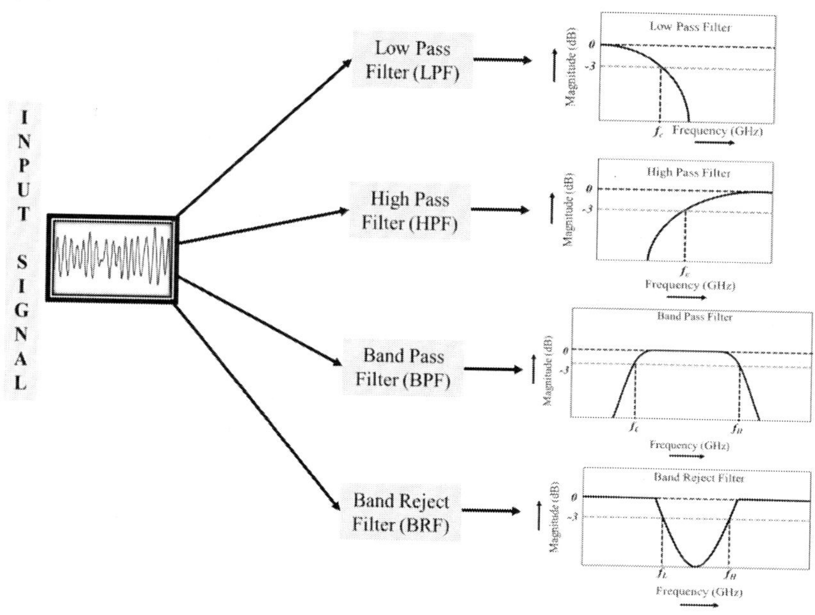

Figure 1. Basic tree diagram along with responses of different filters.

Generally, two approaches are being followed in designing metamaterial-based microwave circuits; (i) the resonant approach where SRRs or complementary SRRs (CSRRs) are more popular. Another one is (ii) the transmission line approach which is based on basic transmission line theory. In order to get broadband devices with less loss, the transmission line approach is mostly chosen as SRR-based circuit heads is lossy. A basic classification of different types of filters is portrayed in Figure 1.

2.1. Resonant Approach

When SRRs (also acts as LC resonant tank circuit) are excited with axial magnetic field, they demonstrate extreme values of magnetic permeability and the same is portrayed in Figure 2(a). SRR based configurations have emerged as a prospective option due to its multifarious application although it exposes negative permeability with narrow frequency range. An array of SRR structures also provide signal rejection capability demonstrating filtering properties when they are properly polarized.

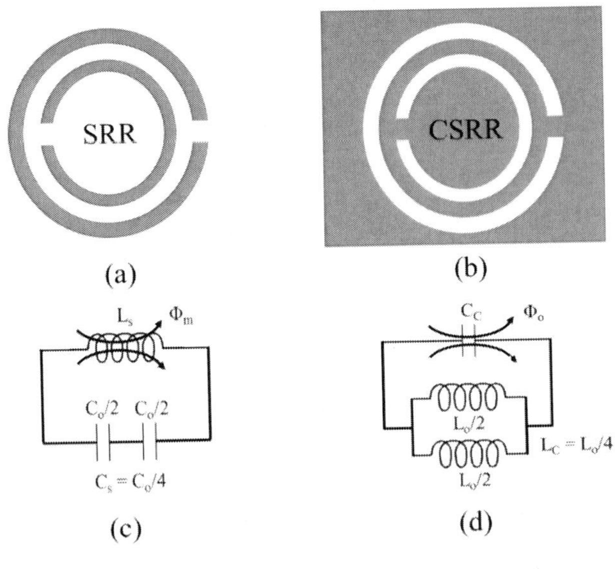

Figure 2. (a) View of Split Ring Resonator (SRR), (b) View of Complementary Split Ring Resonator (CSRR), (c) Equivalent Circuit Model of SRR, and (d) Equivalent Circuit Model of CSRR [10].

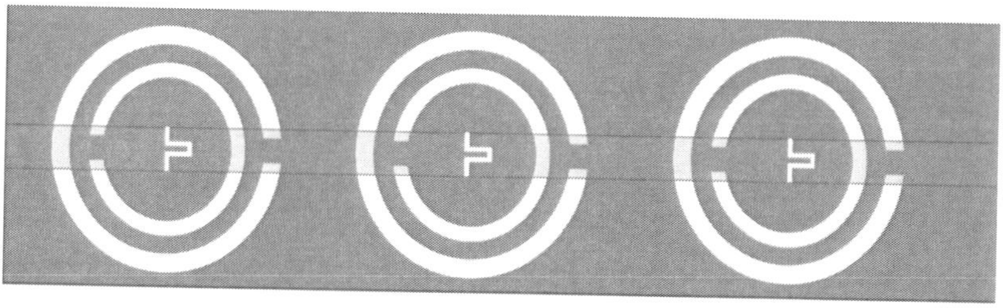

Figure 3. CSRRs along with microstrip line are etched out in ground plane. Capacitive gaps are incorporated on the strip for obtaining a left-handed passband [10].

At the same time, the complementary split-ring resonator (CSRR) structure, which can be used as duality of SRR (by virtue of Babinet principle), where ground plane metallization part is removed for realizing inverse shape of SRR. Therefore, the arrangement scheme for SRRs and CSRRs are preferred to be either top or adjacent or directly beneath of the microstrip lines. Uniform gaps along with incorporation of small elements (may be broadside coupled split-ring resonator (BC SRR) [5]) in the microstrip line help in miniaturizing the filter. The other type of small elements are the spiral resonator (SR), [6-7], multiple CSRRs [8] and square Sirpinski fractal CSRRs [9].

Figure 3 represent SRR, CSRR and their equivalent circuit. A simple left handed (LH) microstrip line with incorporation of CSRRs has been demonstrated in Figure 3 by concept of negative permittivity and the same can be realized using gaps (capacitive in nature) in microstrip line. Due to the negative nature of both ε and μ the same line can be acts as a passband filter.

2.2. Approach Based on Transmission Line (TL) Concept

The theory of transmission lines is already well established and has a predominant relationship with designing and analyzing different microwave circuits. Figure 4 demonstrates homogenous models of right-handed (RH) and left-handed (LH) lossless transmission lines (TL). The concept of implementing low pass filter and high pass filter starts here, and realization is possible by perceiving RH-TL and LH-TL, respectively. The unit cell size, which is considered to be Δz (much smaller than λ_g) and can be approximated as $< \lambda_g/4$ [10].

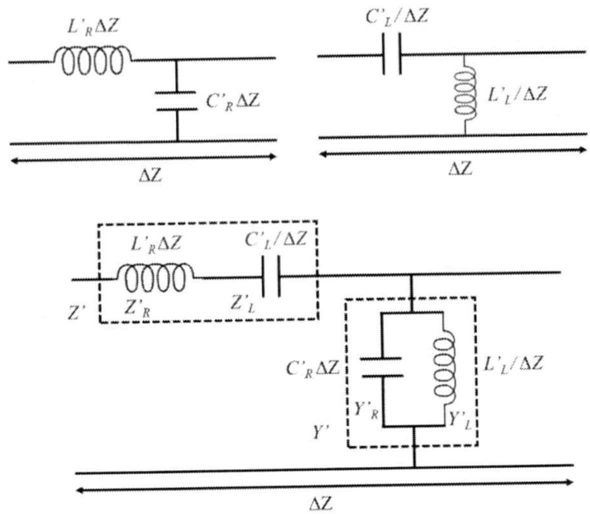

Figure 4. Homogeneous RH TL and LH TL with associated equivalent circuit models [10].

Due to the presence of RH parasitic series inductance and parasitic capacitance, LH TL cannot be realized. Therefore, a composite right and left-handed CRLH TL is the most general representation of the structure. The existence of stop band feature is prominent in the β-diagram of [10] between $\omega_{\Gamma 1}$ and $\omega_{\Gamma 2}$, whereas the same is very difficult in case of pure RH and LH TLs. The shunt and series resonances (equal in nature) can be demonstrated as

$$L'_R C'_L = L'_L C'_R \tag{1}$$

where equal contributions from the RH and LH are observed for balancing each other at given frequency ω_0, this condition can be classified as the *balanced case*. In this context, the concept of quasi-lumped elements [10] is an important factor where the phase response of a TL can be combined with lumped inductance or capacitance or with SRRs and CSRRs.

At a given frequency ω_0, while β tends to zero, the guided wavelength ($\lambda_g = \frac{2\pi}{|\beta|}$) becomes infinite. The extraction of effective parameters has not been further discussed in this chapter as the same has been elaborately discussed in [2, 10].

3. Different Types of Metamaterial Filters

3.1. Left/Right Handed Cells for Metamaterial Based Filters

One of the most adopted approaches in designing transmission line based metamaterial filter is to arrange different LH and RH cells in an alternate way and by combining these cells in a logical fashion in order to get band pass or band stop response. The strategy of obtaining band pass response can be resolved by keeping one RH cell between two LH cells [5-7]. From Figure 5 of [11], it can be depicted that a band-pass behaviour with steep cut-off (both sides of the band) along with poor declining response (upper stop-band) has been nullified by incorporating alternating two types LH and RH transmission line cell. Further, the different bandwidths can also be attained either by coplanar TL based on SRR topology or by CSRR-based configuration [5-8].

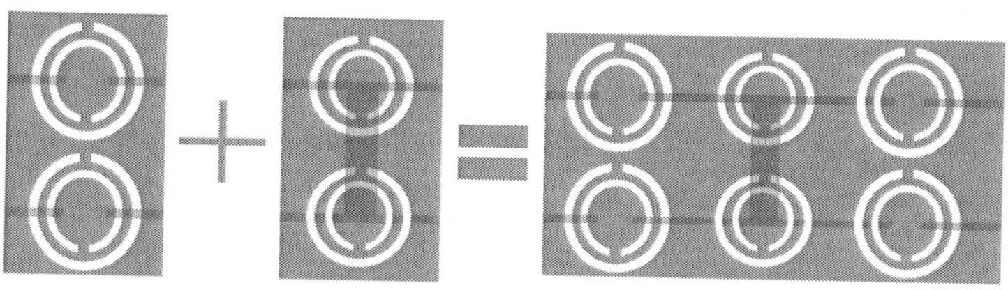

(a) (b) (c)

Figure 5. Layout view of the proposed filter accompanied with (a) CPW LH TL, (b) CPW RH TL, and (c) combination of one RH and two LH unit cells. The thickness of used Arlon substrate is h = 0.49mm and ε_r = 2.43. The smaller SRR radius represented in d is r = 1.39mm, width as well as distance between the rings is 0.2mm (same for all smaller SRRs). The radius of the bigger SRRs is r = 1.52mm. Width between two concentric SRR 2.16mm, and gap length is lg = 1.6mm. The total length of the proposed filter is 1.5cm [11].

In the case of coplanar technology, the combination of SRRs (present in RH lines) is etched out on the bottom side, and capacitive gaps are resemble with CPWs (Figure 5(b) of [11]). On the other hand, CSRR-based microstrip right-handed structures combined with shunt inductances are a good option for obtaining the desired response due to the removal of the metallization part (in the case of CSRRs) from the ground plane. The designed filter in [5] has very selective performance and also provides narrow-band response (symmetric in nature) due to an assemblage of two types of metamaterial unit cells which are coplanar in nature (as shown in Figure 5(c) of [11]). Due to the incorporation of metamaterial unit cell, the length of the filter proposed in [5] has also been reduced (three times shorter) compared to the customary 3rd order Chebyshev filter (coupled-line in nature) with similar responses. An Appendix-I has been incorporated by detailing the metamaterial-based band pass filter design steps.

3.2. UWB Band-Pass Filters Based on Balanced Cells

For designing broadband filters and balanced cells with wider transmission band, composite RH or LH CSRR-based transmission lines are generally used. Some examples, including both kinds (purely resonant and the hybrid cells) of TLs are already discussed in [11]. Adding some additional elements in the transmission line also helps to manage the upper edge of the band due to limited control over the band. An ultra-wide band- combined effect of three balanced unit cells along with three additional CSRRs (which are further formed as big resonator) for covering the standard mask in the case of UWB communications (3.1GHz to 10.6GHz), as depicted in Figure 6(b) of [11]. Three balanced unit cells (starting around 3.1GHz) in combination with additional CSRRs (help to reject the signal around 10.6GHz) are used to achieve high-pass response. This makes the filter to be exposed as pass band behaviour. The same filter (very compact) also covers the desired frequency range. In order to regulate the upper limit of the transmission band along with the inclusion of attenuation poles and suitable rejection of interfering signals (which may appear within the transmission band), the small CSRR-type structures are incorporated as inferred in [12]. Generally, resonators (metallic in nature) are added close to the transmission line on the top layer of the substrate. A similar fashion can be followed on the ground plane either beneath the TLs or inside the bigger resonators if it is complementary in nature. Therefore, spiral resonators (SRs) and complementary spiral resonators (CSRs) approaches have been improvised in the configuration, which is projected in Figure 7(a).

Further, complementary and additional resonators have been etched on the ground plane (illustrated in Figure 7(a)). In this context, incorporating small CSRRs is very promising as it helps nullify the signal level above 10.6GHz and the origination of attenuation pole due to the presence of CSRs [12]. At the same time, these CSRs can also be tuned to achieve desired frequency band and reject undesired signals. In this case, any signal around 4.8GHz has been rejected by using the CSRs.

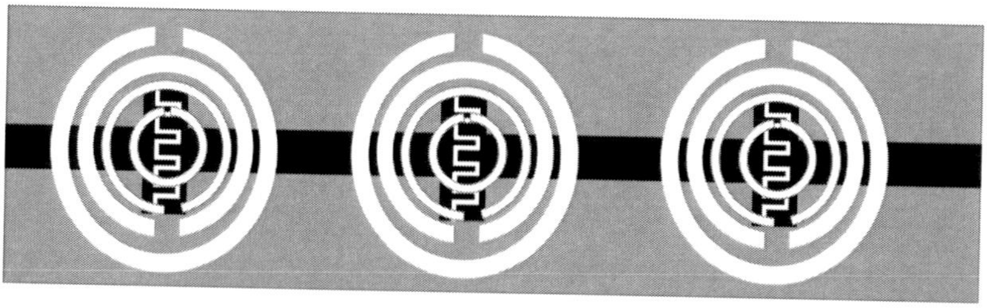

Figure 6. (a) Layout view of a UWBPF filter. CSRRs based balanced cells are used for controlling the upper limit of the band. The used substrate is *Rogers RO3010* ($\varepsilon_r = 10.2$) and thickness is $h = 0.635$mm. For the big SRRs: external ring radius is $r_{ext} = 2.10$mm, $c_{outer} = 0.22$mm and $c_{inner} = 0.16$mm for the external and internal ring, respectively and distance between rings is $d = 0.11$mm. For the small SRRs: external ring radius is $r_{ext} = 0.89$mm, distance between rings is $d = 0.11$mm. Finally $c_{outer} = 0.17$mm and $c_{inner} = 0.17$mm for the external and internal ring, respectively. Interdigital configuration (fingers are separated at a distance of 0.16mm) is also used in order to form capacitor (finger length is 0.84mm and width is 0.14mm) [11].

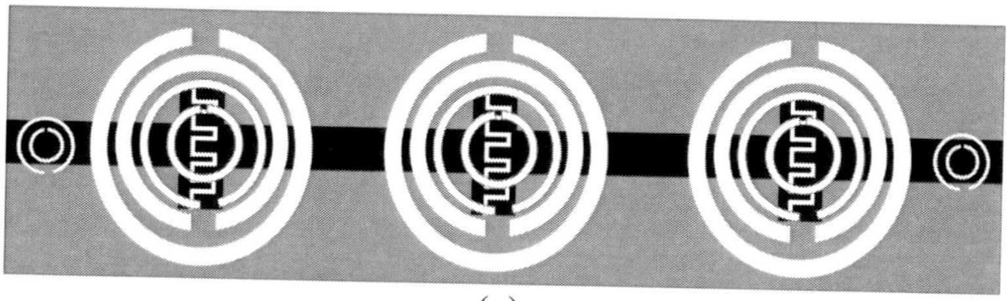

(a)

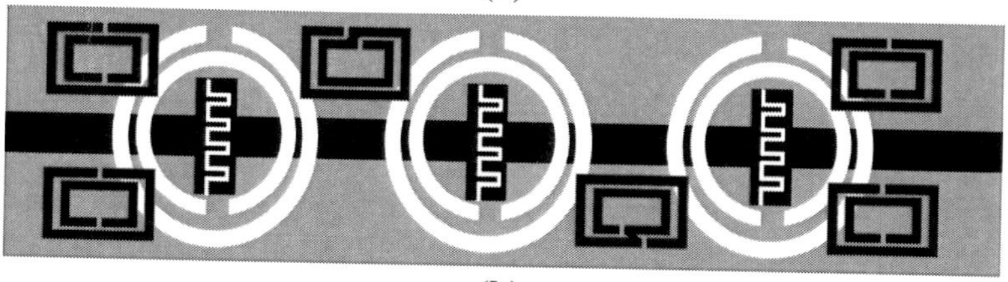

(b)

Figure 7. (a) Layout view and photograph of an additional CSRRs and CSRs loaded UWB filter for controlling the upper limit of the transmission band and to introduce an attenuation pole at 4.8GHz, respectively [12]. The area of the dashed rectangle is $A = 1.77\text{cm} \times 0.41\text{cm}$. (b) Schematic of a SRRs and SRs loaded UWB filter to restrict the transmission band (upper limit) and to introduce an attenuation pole near 5.6GHz, respectively. The area of the rectangle (marked in dashed line) is $A = 1.57\text{cm} \times 0.42\text{cm}$ [11].

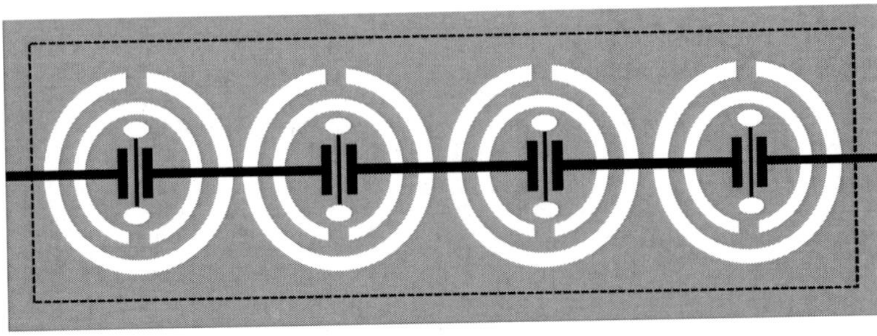

Figure 8. Layout view of the hybrid cell-based UWB filter (total area of the filter < 0.7cm^2). The substrate (*Rogers RO3010)* thickness is $h = 0.127$mm. Line width $W = 0.126$mm, external radius of the outer rings $r = 1.68$mm, and rings are separated with distance $d = 0.19$mm [13].

Figure 7(b) envisages the arrangement of second filter where all the additional metallic resonators are laid on the upper side of the substrate. In this configuration, multiple metallic square resonators are accommodated near to host line, and among them, four SRRs have contributed to filter edges as if they can be used similar to CSRRs. The existence of two SRs near the central part of the filter also influences the attenuation pole at 5.6GHz, and this addition does not contribute to the device's size. Balanced hybrid cells can also be used to implement UWBPFs; the same is presented in Figure 8.

The filter operates from 3.5 to 9GHz with insertion loss (IL) of better than 80dB and the device area is less than 1cm^2. The final device, comprised of four hybrid unit cells, is designated as a balanced configuration to accommodate the desired frequency range. The simulated performance (as portrayed in Figure 11 of [13]) showcases the significant out-of-band rejection response below the S_{21} band satisfying the criteria of desired insertion loss (IL) i.e., IL is better than 80dB at 2GHz. Significant improvement in in-band IL and RL values are observed; simultaneously, no harmonics are observed up to 16GHz. This promising performance is well accommodated within very compact dimensions.

The concept of multimode resonator and stub loaded resonator is also investigated for realizing UBW filters. In [14] concept of quintuple, mode resonator is extended for designing UWB band pass filter based on two transmission zeros allowing it to achieve high selective filter at lower and upper cut-off frequencies. Figure 9 describes the filter layout and measured performance along with ultra-wide stopband behaviour of the realized filter, as projected in Figure 5(b) of [14]. A dual short stub-loaded resonator is constructed in [15] and the same is illustrated in Figure 10.

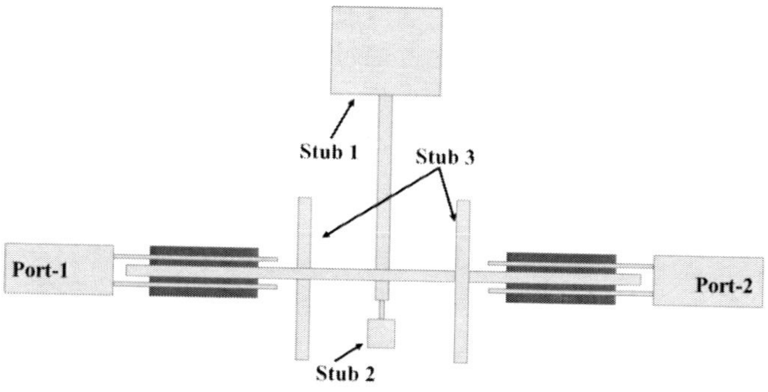

Figure 9. Design of quintuple-mode resonator cum UWB filter [14].

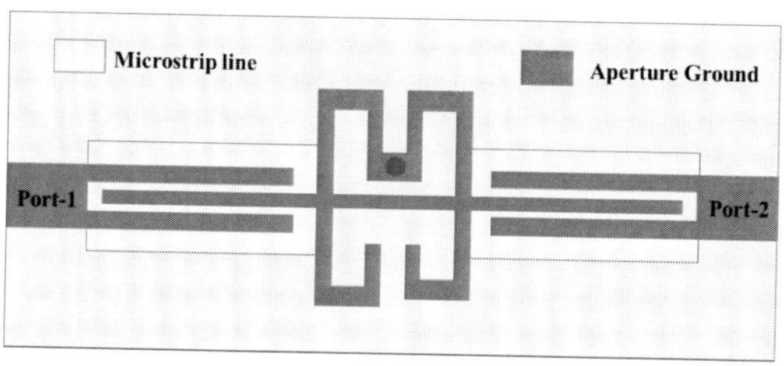

Figure 10. Physical layout view of the quintuple-mode resonator cum UWB filter [15].

In order to analyze the resonant modes of the proposed symmetrical UWB filter, odd-even mode analysis can be used. The measured results of the filter [15] are very promising and exhibit 2.3GHz to 10.4GHz wideband response, and group delay is also within the limit. Simulated and measured results are further compared and presented in Figure 7 [15]. A UWB band pass coupled line filter incorporating defected ground structure is illustrated in [16], ranging from 2.4 to 14.3GHz. The proposed design provides very low insertion loss (0.7dB) at 6.9GHz, and the dimension of the filter is very compact i.e., $26.5 \times 7.54 \times 1.6mm^3$. An interdigital feed line-based UWB BPF with defect (rectangular shaped slots) at the bottom of the ground is introduced, and coupling gaps are provided on the top portion of the substrate. Figure 6 of [16] demonstrates the configuration of the proposed structure, and measured S-parameters are verified with layout performances, as illustrated in Figure 7 of [16]. Another attempt has been made in [17] where a UWB filter provides three notch bands for WiMAX, WLAN and X-band satellite communication applications. The size of the filter $34 \times 12 \times 0.762\text{mm}^3$ which provides 116% fractional bandwidth with low insertion loss 0.4dB from 2.8 to 7.9GHz and 0.8dB from 7.9 to 10.6GHz. The measured performances of the fabricated filter are depicted in Figure 16 and Figure 17 of [17].

3.3. Metamaterial-Based Band-Stop Filters

In this section, different types of metamaterial-based band-stop filter are discussed. Recently Li et al. proposed a miniaturized band-stop filter based on an FSS-based step-impedance resonator (SIR) topology in [18]. A two-dimensional vertical microstrip SIRs are configured in a periodic fashion with reduced thickness. To improve the quasi-elliptic filtering response, a capacitor (in shunt orientation) connecting the centre of SIR has been employed to widen the operating bandwidth. The proposed FSS-based design [18] showcases more than a 40% reduction in thickness. The realized FSS-based band-stop filtering is achieved by selecting a small impedance ratio (of SIR) and tuning the shunt capacitor to provide steep quasi-elliptic band-stop filtering performance. A comparison of simulated and measured results is devised in Figure 9 of [18].

Another periodic structure (I-shaped FSS) is kept inside two rectangular split ring configurations [19] and provides band-stop filtering performance from 9 GHz to 11 GHz. In literature [20], a 3D tunable band-stop FSS is proposed. The overall dimension of the proposed band-stop filter is $3 \times 5 \times 6\text{mm}^3$. A varactor diode-loaded periodic arrangement an annular resonator is presented in [20], which provides a wide frequency tuning range from 3 to 6GHz. Further, two similar configurations are cascaded to improve the bandwidth and selectivity of the band-stop filter. Figure 5 and Figure 7(b) of [20] present the 3D cascaded view of the tunable FSS-based band-stop filter and performance for different incident angles.

3.4. Metamaterial Spatial Filters Based on FSS Structures

3.4.1. Metamaterial Based X-Band Spatial Filter

There is a huge and wide usage of spatial filters, which can be used in a wireless system to isolate any types of disruptions caused by undesired signals in a wireless system from nearby wireless devices. Radio wave transmission using wireless technology has gained wide attention in the modern era without using cables or wires. As the demand for high data rate in mobile communication systems has increased, the option of implementing frequency selective surface (FSS) based spatial filters have also become prime research.

In the class of spatial filter family, exposure of artificial magnetic conductor (AMC) and high impedance surface (HIS) is getting significant attention where impedance bandwidth improvement as well as gain improvement of the antenna/radiator is most common usage, and radar cross section (RCS) is also reduced [21-22]. Planar type of metamaterials, where material thickness is compared to sub-wavelength, can be categorized as meta-surfaces (MSs) or AMC or HIS due to their ease of fabrication [23]. Recently, novel approaches are being followed intensively for designing compact microwave components.

In the electrical domain, FSS based periodic electromagnetic structures behave as spatial filters, either capacitive or inductive shapes [24–32]. High pass, low pass, band pass, and band stop filters can also be designed FSS-based configuration. For converting low pass FSS into a high pass FSS or band pass FSS into band stop FSS, Babinet's theory has proved to be useful [33-35].

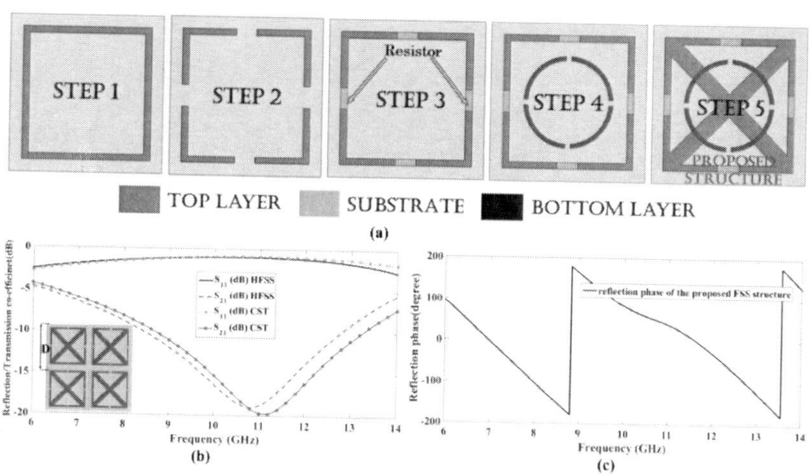

Figure 11. (a) Sequential maturation and design steps of the designed FSS structure, (b) simulated S_{11} and S_{21} characteristics using 3D EM softwares (HFSS and CST), and (c) simulated reflection phase plot of the designed X-band structure (reprinted from [36]).

As seen in Figure 11, the high pass FSS and low pass FSS are both counterparts, and the same concept is also valid for band pass and band stop

FSS. In order to proclaim the equivalent circuit topology of these filters, the RLC circuit can be arranged suitably as per Figure 11.

In general, the FSS patches have the effect of producing resistance (*R*) and inductance (*L*), while the gaps between the FSS have a capacitive effect (*C*). General transmission line theory may be utilized for calculating capacitance (*C*) and inductance (*L*) to explain the physical sense of these *C* and *L* values of various FSS.

In this regard, an example of band stop and band pass filtering has been illustrated in [36]. An ultrathin FSS structure is designed and fabricated for covering the X-band. The design consists of a tetra-arrow, two bi-directional arrows (placed diagonally) at the top, and a circular split metallic structure at the bottom of the low loss RT duroid 5880LZ substrate. The proposed FSS showcase thickness ($0.007\lambda_0$ corresponding to 8.3GHz) of the proposed design is significantly low, and the measured transmission (S_{21}) bandwidth is 47.3% ($S_{21} \leq$ -10dB, 8.15GHz to 13.2GHz). Further, stress on equivalent circuit analysis is also put forward to understand the circuit's insight in a better way.

In the first step, a square loop element (SLE) structure is designed where wide transmission bandwidth is observed over the full X-band. Hence, an SLE has been emerged as a suitable candidate and primarily improvised. Here, unit cell ($L_{unit\text{-}cell}$) size can be approximated using Eq. (3.1).

$$L_{unit-cell} = \frac{c}{2f_r\sqrt{(\varepsilon_s + 1)/2}} \tag{3.1}$$

where, $L_{unit\text{-}cell}$ is unit cell (mm) dimension, c is the speed of light in vacuum (cm/s), f_r is the centre frequency (i.e., 10.65GHz) and is ε_s the dielectric constant of the substrate. The sequential growth of the proposed FSS is portrayed in Figure 11(a), where capacitive effect and inductive effect are combined to form SLE and further modified into a split square structure. In step 2, lumped resistors have been incorporated on the top surface of the structure as the current flows continuously through the gap of the square split structure, making it a square loop element (inductive) [37] and the same is depicted in step 3. Later, in the bottom layer (step 4) circular split type configuration is adopted and a tetra-arrow is reconnected to get wider transmission BW by wisely optimizing the design parameters of the FSS-based configuration.

However, the existence of lumped resistors contributes to ohmic losses apart from the dielectric loss of the substrate. The above factors also help to contribute to the BW broadening of the designed device. High-Frequency Structure Simulator (HFSS) [38] and Computer Simulation Technology (CST) [39] EM packages are used for design and simulation purposes. The proposed FSS structure offers wide simulated transmission (S_{21}) bandwidth, operating from 8.33 to 12.48GHz and near 0 dB reflection over the frequency of interest and results are illustrated in Figure 11(b).

The equivalent circuit model (Figure 12) of the proposed X-band FSS structure is also examined through transmission characteristics, and it has been observed that FSS1 demonstrates band-stop behaviour from 10 to 14GHz, while FSS2 configuration provides band-pass performance from 6 to 14GHz. The combination of FSS1 and FSS2 exhibits the BR response over the desired band (8.15 to 13.2GHz, $S_{21} \leq$ -10dB) and transmission zero located at 10.8GHz. Based on the above consideration, the FSS1 can be framed as a series combination of resistor (R), inductor (L_1) and capacitor (C_1) and can be addressed as a lossy layer due to the appearance of the resistor. The FSS2 (lossless layer) also furnishes band-pass performance, and it can be made compatible within a parallel combination of the inductor (L_2) and capacitor (C_2). As dielectric substrate has been used in order to separate FSS1 and FSS2, therefore it can be modelled as a specific prolonged transmission line (where dielectric spacing is equivalent to the thickness of the substrate (t_{Sub}), and the same is presented in Figure 12(a). From Figure 12(a), the input impedance can be written as [40]

$$Z_{in} = Z_{FSS1} \parallel Z_d \tag{3.2}$$

where,

$$Z_{FSS1} = R + j\omega L_1 + (1/j\omega C_1) \tag{3.3}$$

$$\left.\begin{aligned} Z_{FSS2} &= (j\omega L_2 \parallel 1/j\omega C_2) \\ &= \left(\frac{j\omega L_2}{1-\omega^2 L_2 C_2}\right) \end{aligned}\right\} \tag{3.4}$$

$$Z_d = Z_{sub}\frac{Z_{FSS2} + jZ_{sub}\tan(\beta t_{sub})}{Z_{sub} + jZ_{FSS2}\tan(\beta t_{sub})} \tag{3.5}$$

β is defined as the phase constant for the defined transmission line section $\beta = (2\pi\sqrt{\varepsilon_s} / \lambda,\ Z_{sub} = Z_0 / \sqrt{\varepsilon_s}$ [41-42], and Z_o = 377Ω (free-space impedance). Further, ADS software [43] is used to validate the circuit model and extract the lumped parameters. A 3D circuit simulation and schematic simulation are compared, and the response demonstrates good agreement between these performances (as shown in Figure 12(b). The estimated lumped parameters [43] values are as follows: R = 100ohm, L_1 = 2.099nH, C_1 = 95.04pF, L_2 = 1.049nH, C_2 = 0.569pF.

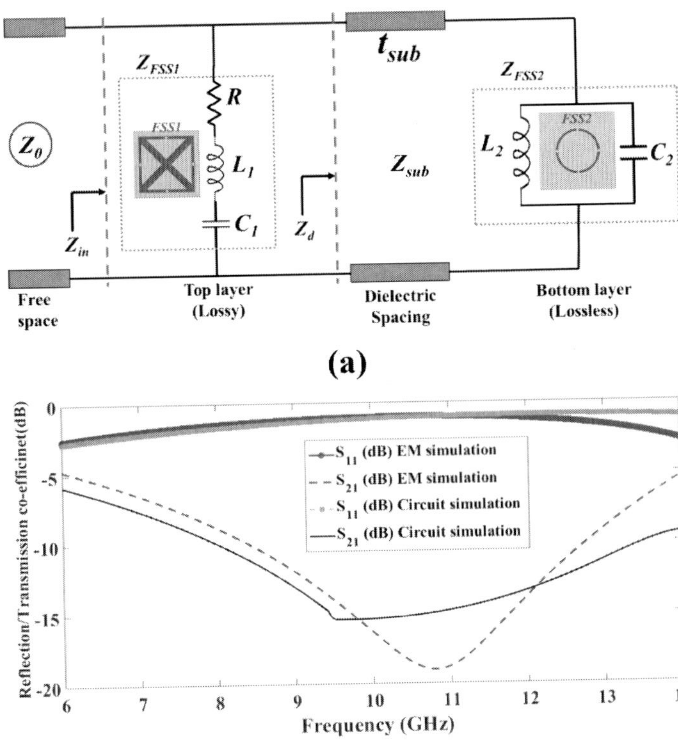

Figure 12. (a) Equivalent circuit model of proposed X-band FSS structure, and (b) reflection (S_{11}) & transmission (S_{21}) characteristics are performed using 3D EM software (HFSS) whereas circuit simulation is performed using ADS software (reprinted from [36]).

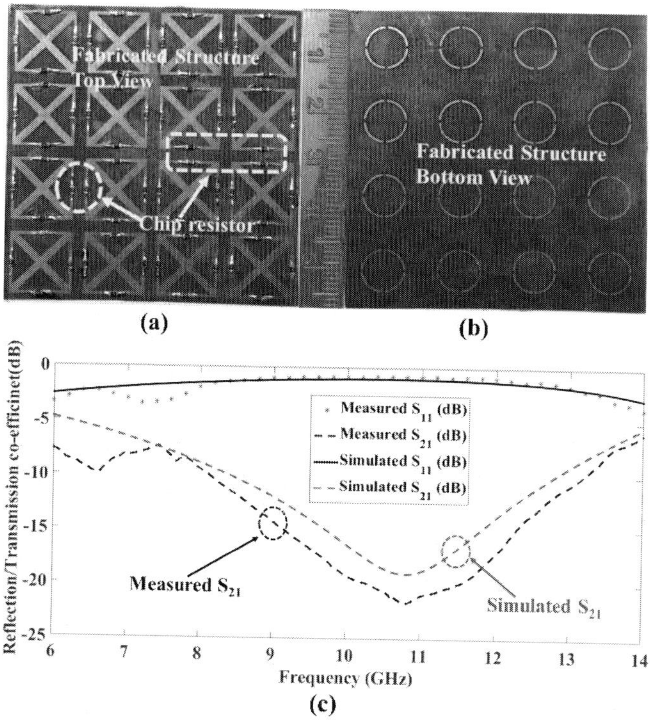

Figure 13. Fabricated X-band FSS structure (a) top layer (FSS1), (b) bottom layer (FSS2), and (c) simulated and measured S_{11} and S_{21} characteristics of the standalone FSS prototype (reprinted from [36]).

Low loss RT duroid 5880LZ substrate is used for fabricating the proposed FSS structure, and metallic patterns (different types) are etched out from the upper side and lower side of the substrate, respectively. Further, chip resistors (surface mount type of value 100ohms, model no. CRCW0603100RFKEA, make: Vishay) are mounted on the top side. The fabricated FSS structure is illustrated in Figures 13(a) (top view) and 13(b) (bottom view). A VNA and a pair of standard gain horn antennas are used for performing S-parameters measurement, where two antennas are used (one transmitting the EM wave and the other receiving the transmitted signal). The transmission (S_{21}) BW of the proposed structure is 47.3% (measured BW and covers 8.15 to 13.2GHz), and a comparison is presented in Figure 13(c). The proposed X-band FSS structure renders 47.3% measured wide transmission (S_{21}) bandwidth and emerged as an ultra-thin FSS structure with an electrical thickness of $0.007\lambda_0$ only, corresponding to the lowest operating frequency.

3.4.2. Metamaterial Based Ultra-Wideband (UWB) Spatial Filter

Another attempt has been made and presented in [44], where the proposed FSS structure can be used as a band-stop spatial filter. The same FSS structure can be decomposed into band-pass and band-stop filters based on their structural orientation, and the same is discussed in the subsequent paragraph.

In literature, some basic formula (3.6) is available for estimating the approximate unit cell ($L_{unit-cell}$) size of periodic structure and is provided below as

$$L_{unit-cell} = \frac{c}{4f_r\sqrt{(\varepsilon_s + 1)/2}} \tag{3.6}$$

where unit cell ($L_{unit-cell}$) size is in mm, c is the speed of light in vacuum (cm/s), f_r is calculated corresponding to 3.5GHz (lowest transmission frequency) and dielectric constant (of the substrate) is defined as ε_s. The proposed FSS structure has two layers, and layers are separated by foam (thickness = t_{foam} = 4mm) substrate. FSS1 (positioned on the top side of the substrate) is configured as a square patch integrated with "+ *sign*", and rotated at an angle of 45°, whereas only square patch is adopted in the bottom layer, denoted as FSS2. Figure 14(a) to Figure 14(c) represent the top view (unit cell), bottom view (unit cell) and side view of the designed FSS and design parameters are illustrated in the caption of Figure 14. RT duroid 5880LZ (low loss) substrate (ε_s = 2.2) is used for designing the dual-layer FSS structure. The unit cell dimension is 15mm and fabricated with thickness (t_{sub1} = t_{sub2} = t_{sub}) of 0.254mm. Numerical simulation is carried out using equations available in [38-39]. By suitably arranging the elements and inter-element spacing of different shapes, wide reflection or transmission bandwidth can be achieved. Therefore, the necessity of compact low-profile UWB periodic structures is getting attention day by day due to its impressive response in space communication and aircrafts technology. Hence, square loop [41] configuration has been chosen widely designing FSS-based spatial filters where BP or BS filtering scheme can be accommodated wisely with required reflection (S_{11})/transmission (S_{21}) characteristics. Moreover, SLE prevails smaller gap between inter elements compared to patch type element making it suitable for providing wide S_{21} or S_{11} characteristics within the operating frequency range [45-47].

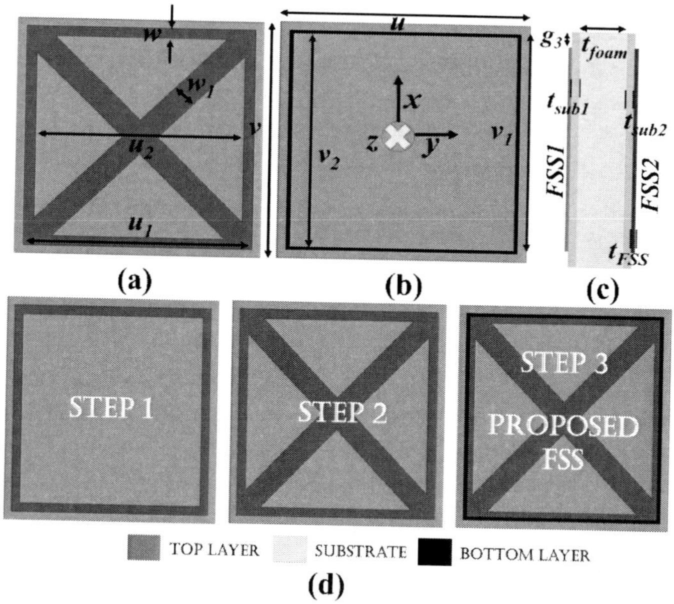

Figure 14. Proposed UWB-FSS structure (a) top view, (b) bottom view, (c) side view, and (d) sequential evolution and design steps. $u = v = 15$, $u_1 = 13.5$, $u_2 = 12.3$, $v_1 = 13.5$, $v_2 = 13$, $w = 0.6$, $w_1 = 1.5$, $w_2 = 0.4$, $g_2 = 0.5$, $g_3 = 0.5$, $r_{FSS} = 3.1$ (dimensions are in mm) (reprinted from [44]).

The unit cell dimension primarily considered is calculated from equation (3.6), and the shape is transformed into tetra arrow configuration (which also possesses four-fold symmetry) to conserve the desired transmission BW. A dual layer structure (which can provide multiple resonances) is exploited here to attain larger bandwidth. The sequential evolution and development steps of the proposed UWB FSS structure are demonstrated in Figure 14(d). The impression of achieving wide targeted transmission (S_{21}) BW has destined for designing such two-layer FSS geometry. The combined effect of FSS1 and FSS2 (Figure 14(c)) provides wider transmission (S_{21}) BW from 4 to 12.25GHz, whereas standalone FSS1 and FSS2 exhibit relatively smaller S_{21} bandwidth (simulated transmission BW from 5.81 to 12.19GHz for FSS1 and 3.44 to 5.13GHz for FSS2).

The detailed analysis of FSS structure is further analyzed from physics point of view in this section. If the inter-element spacing length (ISL) of the FSS unit cell (in the case of our design, ISL = 15mm) is smaller and approximately equal to wavelength (corresponding to operating frequency), then analysis of such unit cell can be included in resonance region [41].

The same criteria are also satisfied in our design, where the operating wavelength (λ_g) is 19.08mm (corresponding to 10.6GHz and ISL is < λ_g) [41, 47-49]. For pointing out band-pass/band-stop behaviour, individual FSS layer (FSS1 and FSS2) performances are observed, and both the layers demonstrate BS responses corresponding to standalone S_{21} responses ($S_{21} \leq$ -10dB BW for FSS1: 3.44 to 5.13GHz and for FSS2: 5.81 to 12.19GHz) which is portrayed in Figure 15(a). Therefore, the proposed UWB-FSS structure can be portrayed as a series combination of the inductor (L_1), and a capacitor (C_1) (as an equivalent circuit model). A certain length of TL isolates the two-layer FSS. The circuit model with all elements (L and C), is shown in Figure 15(b) and can be correlated with Figure 14(c) also.

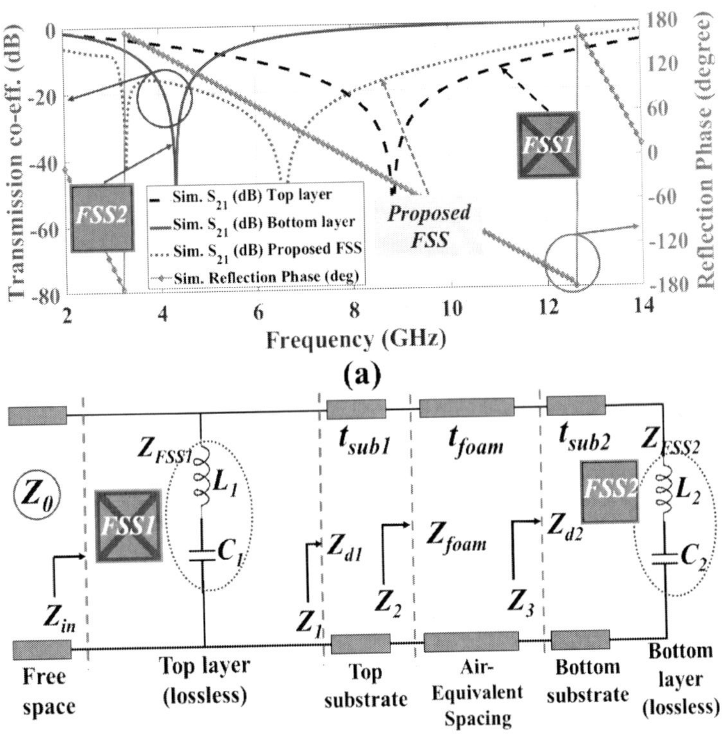

(b)

Figure 15. (a) Simulated transmission (S_{21}) characteristics of individual FSS along with stacked FSS structure; simulated reflection phase is projected in the right y-axis, and (b) circuit layout of the designed UWB-FSS structure (reprinted from [44]).

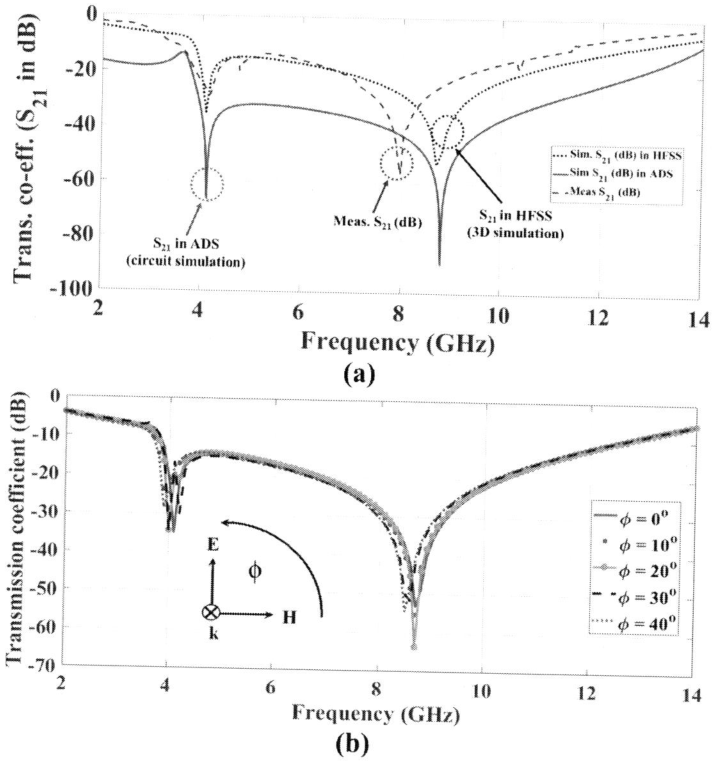

Figure 16. (a) Simulated and measured S_{21} characteristics of the proposed UWB-FSS structure are compared, and (b) simulated transmission characteristics for different polarization angles is projected (for different normal incidence) (reprinted from [44]).

Total input impedance of the equivalent circuit model can also be written as

$$Z_{in} = Z_{FSS1} \parallel Z_1 \tag{3.7}$$

$$\text{Now, } Z_{FSS1} = (j\omega L_1 + 1/j\omega C_1) \tag{3.8}$$

$$Z_{FSS2} = (j\omega L_2 + 1/j\omega C_2) \tag{3.9}$$

Input impedance at each section of the transmission line can also be estimated as (refer to Figure 15(b))

$$Z_3 = Z_{d2}\frac{Z_{FSS2} + jZ_{d2}\tan(\beta_2 t_{sub2})}{Z_{d2} + jZ_{FSS2}\tan(\beta_2 t_{sub2})} \quad (3.10)$$

$$Z_2 = Z_{foam}\frac{Z_3 + jZ_{foam}\tan(\beta_{foam} t_{foam})}{Z_{foam} + jZ_3\tan(\beta_{foam} t_{sub2})} \quad (3.11)$$

$$Z_1 = Z_{d1}\frac{Z_2 + jZ_{d1}\tan(\beta_1 t_{sub1})}{Z_{d1} + jZ_2\tan(\beta_1 t_{sub1})} \quad (3.12)$$

where, $Z_{di} = Z_0/\sqrt{\varepsilon_{si}}$, $\beta_i = 2\pi\sqrt{\varepsilon_{si}}/\lambda$, i = 1, 2. Z_i and β_i are the intrinsic impedance and phase constant of ith transmission line section [47-48] respectively as detailed in Figure 15(b). As the substrate thickness (t_{sub1} and t_{sub2}) for FSS1 and FSS2 are equal, then automatically Z_{d1} and Z_{d2} comes out as equal too. Further, circuit validation is performed by using ADS software [43]. Simulated transmission characteristics, performed by using HFSS and ADS, is illustrated in Figure 16(a)). The extracted lumped parameters values are (using ADS [43]) enlisted below: L_1 = 0.739nH, L_2 = 0.471nH, C_1 = 2.014pF and C_2 = 0.695pF. Steps for extracting the values of L and C has also been detailed and appended in the Appendix-II.

The proposed UWB-FSS is fabricated (using low loss RT duroid 5880LZ) substrate thickness is only 0.254mm, which is ultra-thin in nature. Vector Network Analyzer (VNA) is used for performing S-parameters measurement. The measured S_{11} (dB) and S_{21} (dB) responses of the fabricated FSS is presented in Figure 16(a) which demonstrates 108% transmission ($S_{21} \leq$ -10dB) bandwidth from 3.5GHz to 11.75GHz.

Conclusion

This chapter emphasizes the detailed review on design approachs of different types of filters based on metamaterial topology. Design approaches for different types of resonant configuration is adopted and discussed with specific examples. In some of the portion, diverse design philosophy starting from band pass, band stop, and spatial filters with band stop as well as UWB topology is well elaborated along with design examples from different research group. The performance of these filters is also illustrated and at the same time incorporation of metamaterial also helps to reduce the size of these filters significantly. Moreover, design steps of metamaterial based filter as well as spatial filter is also established in appendix-I & II. Overall the chapter destines the use of possible metamaterial structures and can be adopted as good alternative solution in designing compact, UWB filters. Finally a comparison table is appended where detailed comparison of existing metamaterial based structure is illustrated with the proposed spatial filter (band stop and UWB filter).

Table 1. Comparison of the proposed X-band and UWB spatial filter with other available literatures

Ref. No.	**Dimension of FSS or metamaterial unit cell along with type of FSS**	**FSS thickness (at lowest frequency)**	**Frequency of operation and S_{21} BW**
[50]	12mm, split ring resonator (SRR) configuration	$0.016\lambda_0$, @6GHz	2.9GHz to 6.41GHz (75.4%) and 2.6GHz to 6.6GHz (87%)
[51]	L-shaped unit cell configuration	$0.032\lambda_0$, @6GHz	0.48GHz to 6.5GHz, 172.49%
[52]	7mm, patch type and grid type aperture	$0.027\lambda_0$, @5GHz	5.2GHz to 8.3GHz, 45.9%
[53]	9.8mm, Asymmetric V-shaped configuration	$0.025\lambda_0$, @2.4GHz	2.18GHz to 3.04GHz, 32.9%
[54]	7mm, cross-shaped slots (4 nos.) and a MTM based ring slot	$0.033\lambda_0$, @5.01GHz	5.01GHz to 7.35GHz, 37.9%
[55]	10.6mm and split ring configuration (top layer), 10.8 mm and Jerusalem cross configuration (bottom layer)	$0.074\lambda_0$, @3.1GHz	3.1GHz to 10.5GHz, 105.3%
[56]	44mm, planar patch type configuration in upper and bottom layer	$0.011\lambda_0$, @2.1GHz	1.7GHz to 2.2GHz, 25.6%
Proposed structure1 (X-band spatial Filter)	14.5mm, tetra-arrow configuration (lumped resistor loaded) and circular split configuration at top and bottom layer respectively	$0.007\lambda_0$, @8.3GHz	8.15GHz to 13.2GHz, 47.3%
Proposed structure1 (UWB spatial Filter)	15mm, tetra-arrow configuration at top layer whereas at rectangular split configuration is adopted at bottom layer	$0.029\lambda_0$, @3.5GHz	3.5GHz to 11.75GHz, 108%

Appendix-I

A generalized filter design steps are illustrated in this appendix-I. For designing BPF or BRF resonator length and microstrip configuration (parallel line coupled, hairpin resonator, comb-line configuration, open stub resonator etc.) can be selected and chosen based on the frequency of operations. Section 4.1.5 of [57] can be used to calculate resonator length and guided wavelength. In case of narrow bandwidth operation, slow wave open-loop resonators can be accommodated, and the analysis steps can be determined from section 11.4.3 of [57], where it is important to determine the mutual coupling, element values corresponding to the order of the filter and other design parameters (length and width of resonators as well as the gap between two consecutive resonators). The detailed example of different filters along with estimated parameters are properly cited in [57].

Appendix-II

Different methods can be adopted for estimating equivalent inductance (L) and capacitance (C) of any structure, available in open literature. Investigation of surface current distribution in FSS [42] and curve fitting technique [49] are one of the easy solution for the same. Here, curve fitting technique is used considering its feasibility, easy handling and reliability. Step by step procedures of the curve fitting technique are described below.

Step 1: First step is the identification of BS or BP behaviour of any periodic structure which can be concluded from its S_{21} or S_{11} characteristics.

In this chapter the behaviour of such individual FSS structures has been analyzed and transmission and reflection responses are observed. Both the FSS1 and FSS2 configurations provide BS behaviour i.e for FSS1 $S_{21} \leq -10$ dB BW is 3.44 to 5.13GHz and for FSS2: 5.81 to 12.19GHz. This phenomenon is demonstrated in Figure 14(a).

Step 2: Framing of an equivalent circuit of the designed FSS structure (layer-wise) has been stacked by applying the transmission line (TL) theory [40]. The BS performance for both the layers of FSS (FSS1 and FSS2) are arranged as a series combination of the inductor (*L*) and capacitor (*C*) (displayed in Figure 15(b)).

Step 3: In step 3, initial *L* and *C* values can also be estimated from Eqn. (8.38) in [40], and those are considered here for modelling the equivalent circuit performing circuit-level simulation using ADS software (also by modelling with a transmission line of a specific length). These steps are illustrated in Figure 17(a).

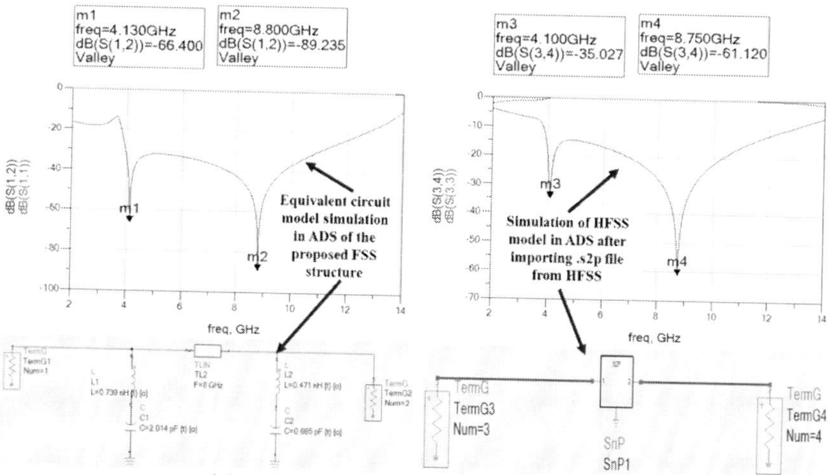

Figure 17. (a) ADS simulation is performed accommodating lumped parameters along with response of the proposed FSS structure, and (b) schematic simulation of equivalent HFSS model using ADS, after importing the .s2p file of the designed FSS from HFSS (reprinted from [44]).

Step 4: Further, the optimization and tuning are carried out for L_1, C_1, L_2 and C_2 until the desired performance (transmission/reflection characteristics) is achieved. In this regard, one can set an optimization goal in ADS tool for desired response in accordance with the simulation results, obtained in HFSS software. For that, designer has to export .s2p file from HFSS and define the same in ADS schematic, as shown in Figure 17(b).

Step 5: Finally, the 'Goal' can be set according to the desired response and optimization can be carried out.

References

[1] Martin, F., F. Falcone, J. Bonache, R. Marques, and M. Sorolla, "Split ring resonator based left handed coplanar waveguide," in *Applied Physics Letters*, vol. 83, pp. 4652–4654, 2003.

[2] Deyasi, A., P. Debnath, A. K. Datta, and S. Bhattacharyya, "*Photonics, Plasmonics and Information Optics*," CRC press, 2021.

[3] Garcia-Garia, J., J. Bonache, F. Falcone, J. D. Baena, F. Martin, I. Gil, T. Lopetegi, M. A. G. Laso, A. Marcotegui, R. Marques, and M. Sorolla, "Stepped-impedance low pass filters with spurious passband suppression," in *Electronics Letters*, vol. 40, pp. 881–883, 2004.

[4] Ali, A., M. A. Khan, and Z. Hu, "High selectivity low-pass filter using negative-e metamaterial resonators," in *Electronics Letters*, vol. 43, pp. 528–530, 2007.

[5] Marques, R., F. Medina, and R. R.-E. Idrissi, "Role of bianisotropy in negative permeability and left handed metamaterials", in *Phys. Rev. B, Condens. Matter*, vol. 65, pp. 14441–14446, Apr. 2002.

[6] Baena, J., R. Marques, F. Medina, and J. Martel, "Artificial magnetic metamaterial design by using spiral resonators", in *Phys. Rev. B, Condens. Matter*, vol. 69, pp. 14402–14402, Jan. 2004.

[7] Falcone, F., F. Martin, J. Bonache, M. A. G. Laso, J. Garcia-Garcia, J. D. Baena, R. Marques, and M. Sorolla, "Stopband and band pass characteristics in coplanar waveguides coupled to spiral resonators", in *Microwave Opt. Technol. Lett.*, vol. 42, pp. 386–388, Sep. 2004.

[8] Bengin, V. C., V. Radonic and B. Jokanovic, "Left-handed microstrip lines with multiple complementary split-ring and spiral resonators", in *Microwave and Optical Tech. Letters*, vol. 49, no. 6, pp. 1391-1395, June 2007.

[9] Bengin, V. C., V. Radonic and B. Jokanovic, "Fractal Geometries of Complementary Split-Ring-Resonators", in *IEEE MTT Trans.*, vol. 56, no. 10, pp. 2312-2321, Oct. 2008.

[10] Caloz, C. and T. Itoh, "*Electromagnetic Metamaterials: Transmission Line Theory and Microwave Applications*," Wiley Publication, 2006.

[11] Gil, M., J. Bonache, and F. Martin, "Metamaterial Filters: A review," *Science Direct,* Elsevier, pp. 186-197, 2008.

[12] Martin, F., M. Gil, J. Bonache, and J. Garcia-Garcia, "Composite Right/Left Handed (CRLH) Transmission Lines Based on Complementary Split Rings Resonators (CSRRs) and Applications," in *Proceedings of the International Congress on Advanced Electromagnetic Materials in Microwaves and Optics (Metamaterials 2007)*, Rome, Italy, pp. 872–875, Oct. 2007.

[13] Gil, M., J. Bonache, J. Garcia, J. Martel, F. Martin, "Composite Right/Left Handed Metamaterial Transmission Lines Based on Complementary Split Rings Resonators and their Applications to Very Wideband and Compact Filter Design," in *IEEE MTT Trans.*, vol. 55, pp. 1296–1304, Oct. 2007.

[14] Wu. X. H., Q. Chu, X. Tian, X. Ouyang. "Quintuple-mode UWB band-pass filter with sharp roll-off and super-wide upper stopband," in *IEEE Microwave and Wireless Components Lett.*, vol. 21, pp. 661-663, 2011.

[15] Zhu, H. and Q. X. Chu, "Compact ultrawideband (UWB) band-pass filter using dual-stub-loaded resonator (DSLR)," in *IEEE Microwave and Wireless Components Lett.*, vol. 23, pp. 527-529, 2013.

[16] Ansal, K., D. S. Jose, A. S. Kumar, and T. Shanmugantham, "A band pass coupled line filter with DGS for ultra-wide band application," in *Third International*

Conference on Computing and Network Communications (CoCoNet-19), vol. 171, pp. 561-567, 2020.

[17] Basit, A., M. I. Khattak, and M. Al-Hasan, "Design and analysis of a microstrip planar UWB band-pass filter for WiMax, WLAN, and X-band satellite communication systems," in *Progress In Electromagnetic Research M*, vol. 93, pp. 155-164, 2020.

[18] Li, B. and Z. Shen, "Miniaturized bandstop frequency-selective structure using stepped-impedance resonators," in *IEEE Antennas and Wireless Propagation Lett.*, vol. 11, pp. 1112-1115, 2012.

[19] Sahai, J. and G. Tiwari, "A band pass coupled line filter with DGS for ultra-wide band application," in *Third International Conference on Computing and Network Communications (CoCoNet-19)*, vol. 171, pp. 561-567, 2020.

[20] Jia, S., B. Xu, and T. Zheng, "Design of a 3-D tunable band-stop frequency selective surface with tuning range," in *Progress In Electromagnetic Research M*, vol. 92, pp. 9-16, 2020.

[21] Munk, B. A. *"Frequency Selective Surfaces: Theory and Design,"* Wiley, New York, NY, USA (2007).

[22] Kushwaha, A., and R. Kumar, "Design of slotted ground hexagonal microstrip patch antenna and gain improvement with FSS screen," in *Progress In Electromagnetic Res. B.*, vol. 51, pp. 177-199, 2013.

[23] Anwar, R. S., L. Mao, and H. Ning, Frequency Selective Surfaces: A Review, *Applied Sciences*, vol. 8, app8091689, 2018.

[24] Niroo-Jazi, M., T. A. Denidni, Electronically sweeping-beam antenna using a new cylindrical frequency-selective surface, in *IEEE Antennas and Wireless Propagation Lett.* vol. 12, pp. 666–676, 2013.

[25] Kapoor, A., K. R. Jha, G. Singh, "Design and analysis of frequency selective surface at Ka/K/Ku Band for antenna miniaturization," in *2016 International Conference on Computational Techniques in Information and Communication Technologies (ICCTICT)*, pp. 25–31, 2016.

[26] Zhao, P., Y. Zhang, R. Sun, W. Zhao, Y. Hu and G. Wang, "Design of a novel miniaturized frequency selective surface based on 2.5-Dimensional Jerusalem cross for 5G applications, *Wireless Communications and Mobile Computing*," vol. 2018, Article 6 pages, 2018.

[27] Liu, H. L., K. L. Ford, R. J. Langley, "Design methodology for a miniaturized frequency selective surface using lumped reactive components," in *IEEE Transactions on Antennas and Propagation*, vol. 57, pp. 2732–2738, 2009.

[28] Agarwal, K. Nasimuddin and A. Alphones, "Unidirectional wideband circularly polarized aperture antennas backed with artificial magnetic conductor reflectors," in *IET Microwaves, Antennas & Propagation*, vol. 7, pp. 338–346, 2013.

[29] Moharamzadeh, E., A. M. Javan, "Triple-band frequency selective surfaces to enhance gain of X-Band triangle slot antenna," in *IEEE Antennas and Wireless Propagation Letters*, vol. 12, pp. 1145–1148, 2013.

[30] Cure, D., T. M. Weller, T. Price, F. A. Miranda, F. W. Van Keuls, "Low-profile tunable dipole antenna using barium strontium titanate varactors," in *IEEE Transactions on Antennas and Propagation,* vol. 62, pp. 1185–1193, 2014.

[31] Genovesi, S., F. Costa, A. Monorchio, "Wideband radar cross section reduction of slot antennas arrays," in *IEEE Transactions on Antennas and Propagation*, vol. 62, pp. 163–173, 2014.

[32] Anwar, R., Y. Wei, L. Mao, and H. Ning, "Miniaturised frequency selective surface based on fractal arrays with square slots for enhanced bandwidth," in *IET Microwaves, Antennas & Propagation*, vol. 13, pp. 1811–1819, 2019.

[33] Ulrich, R. "Far-infrared properties of metallic mesh and its complementary structure," in *Infrared Physics*, vol. 7, pp. 37-55, 1967.

[34] Lee, S. W., G. Zarrillo, and C. L. Law, "Simple formulas for transmission through periodic metal grids or plates," in *IEEE Transactions on Antennas and Propagation*, vol. 30, pp. 904–909, 1982.

[35] Mittra, R., C. H. Chan, and T. Cwik, "Techniques for analyzing frequency selective surfaces-A review," in *Proceeding IEEE*, vol. 76, pp. 1593–1615, 1988.

[36] Bhattacharya, Anamiya, Bidisha Dasgupta, and Rajeev Jyoti, "Design and Analysis of Ultrathin X-Band Frequency Selective Surface Structure for Gain Enhancement of Hybrid Antenna", in *International Journal of RF-Microwave Computer-Aided Engineering*, December 2020.

[37] Xu, N., J. Gao, J. Zhao, and X. Feng, "A novel wideband, low-profile and second-order miniaturized band pass frequency selective surfaces," *AIP Advances*, vol. 5(077157), 2015.

[38] *Ansys High Frequency Structural Simulator (HFSS)*, Version: 18.2.

[39] *Computer Simulation Technology (CST)*, Version: 2019.

[40] Pozar, D. M. "*Microwave Engineering,*" Wiley publication, 4th edition, 2013.

[41] Costa, F., A. Monorchio, and G. Manara, "Efficient analysis of Frequency-Selective Surface by a simple equivalent circuit model," in *IEEE Antennas and Propag. Mag.*, vol. 54, pp. 35-48, 2012.

[42] Ghosh, S. and K. V. Srivastava, "An equivalent circuit model of FSS-based metamaterial absorber using coupled line theory," in *IEEE Antennas and Wireless Propag Lett.*, vol. 14, pp. 511-514, 2015.

[43] *Advanced Design System* (ADS), Version: 2019.

[44] Bhattacharya, Anamiya, Bidisha Dasgupta, and Rajeev Jyoti, "A Simple FSS Structure for Performance Improvement UWB Antenna in Frequency and Time Domains," in *International Journal of RF-Microwave Computer-Aided Engineering*, August 2021.

[45] Xu, Y. and M. He, "Design of multilayer frequency selective surfaces by equivalent circuit method and basic building blocks," *Intl. Journal of antennas and Propagation*, (2019), Article ID: 9582564.

[46] Sheng, X. J., J. J. Fan, N. Liu, and C. B. Zhang, "A miniaturized dual-band FSS with controllable frequency resonances," *IEEE Microwave and Wireless Components Lett.*, 17, pp. 915–917, 2017.

[47] Chatterjee, A. and S. K. Parui, "A triple-layer dual-bandpass frequency selective surface of third order response with equivalent circuit analysis," *Int. J RF Microw. Comput. Aided Eng.*, e22047, 2020.

[48] Ghosh, S., S. Bhattacharyya, and K. V. Srivastava, "Design, characterization and fabrication of a broadband polarization-insensitive multilayer circuit analogue absorber," *IET Microw. Antennas Propag.*, Vol. 10, pp. 850-855, 2016.

[49] Liu, N., X. Sheng, C. Zhang, J. Fan, and D. Guo, "A design method for synthesizing wideband band-stop FSS via its equivalent circuit model," *IEEE Antennas and Wireless Propagation Letters*, vol. 16, pp. 2721-2725, 2017.

[50] Alibakhshikenari, M., M. N. Moghadasi, and R. A. Sadeghzadah, Bandwidth and radiation specifications enhancement of monopole antennas loaded with split ring resonators, *IET Microw. Antennas Propag.*, 14, pp. 1487–1496, 2015.

[51] Alibakhshikenari, M., B. S. Virdee, A. Ali, and E. Limiti, Miniaturised planar-patch antenna based on metamaterial L-shaped unit-cells for broadband portable microwave devices and multiband wireless communication systems, *IET Microw. Antennas Propag.*, 12, Iss. 7, pp. 1080-1086, 2018.

[52] Chatterjee, A. and S. K. Parui, "Gain enhancement of a wide slot antenna using a second-order band-pass frequency selective surface," *Journal of Radio engineering*, Vol. 24, pp. 455-461, 2015.

[53] Ameen, M. and R. K. Chaudhary, Metamaterial-based wideband circularly polarised antenna with rotated V-shaped metasurface for small satellite applications, *Electron. Lett.*, 55, No. 7, pp. 365–366, 2019.

[54] Li, X., Y. C. Jiao, and L. Zhang, Wideband low-profile CPW-fed slot-loop antenna using an artificial magnetic conductor, *Electron. Lett.*, 54, No. 11, pp. 673–674, 2017.

[55] Kundu, S., A. Chatterjee, S. K. Jana, and S. K. Parui, "A compact umbrella-shaped UWB antenna with gain augmentation using frequency selective surface," *Journal of Radio engineering*, Vol. 27, pp. 448-454, 2018.

[56] Dong, F., L. Xu, W. Lin, and T. Zhang, A Compact Wide-Band Hybrid Dielectric Resonator Antenna with Enhanced Gain and Low Cross-Polarization, *Intl. Journal of antennas and Propagation*, Article ID: 6290539, 2017.

[57] Hong, J. S. and M. J. Lancaster, *Microstrip filters for RF/Microwave applications*, John Wiley & Sons, Inc., 2001.

Chapter 2

Metamaterial Inspired Advanced Radio Frequency Identification and Applications

Durga Prasad Mishra and Santanu Kumar Behera*
Department of Electronics and Communication Engineering
National Institute of Technology, Odisha, India

Abstract

Metamaterial (MTM) science has been a developing multidisciplinary field in electromagnetics and antenna engineering research for decades. The intriguing properties of metamaterials sparked a flurry of studies in diverse disciplines, such as radio frequency identification (RFID) technology, microwave devices, antennas, optical communication, etc. MTM structures show abnormal electromagnetic properties which are not readily available naturally. These properties include negative permittivity, negative permeability, wide alternation in phase, inversion in Doppler's effect, and Snell's law. These exclusive electromagnetic characteristics of materials give rise to a new era in RFID technology, covering widespread usage in retail, healthcare, and tracking domains. The split-ring resonator (SRR), an evolutionary structure in antenna design introduced in the nineties, exhibits compact size and strong negative permittivity. This chapter aims to provide a brief idea about metamaterial-inspired structural design and analysis, focusing on circular and rectangular SRRs for RFID applications. The important features like radar cross-section (RCS), read range, radiation resistance, the surface current distribution, and the coding capacity are analyzed, and related literature is presented, covering a wide range of applications by showing the importance of metamaterials in the emerging era of advanced

* Corresponding Author's Email: skbehera@ieee.org.

In: Metamaterials for Microwave and Terahertz Applications
Editors: B. Appasani, O. Prakash Acharya, A. Vidyakant Jha et al.
ISBN: 979-8-88697-330-3

communication technology. The chapter also discusses the implementational challenges, including real-time problems that can help find a new research direction in metamaterial structural science.

Keywords: chipless radio frequency identification, metamaterial, split-ring resonator, radar cross section

1. Introduction

Metamaterials (MTMs) and meta-surfaces play a vibrant role in improving the working range and power transfer efficiency of wireless power transfer (WPT) systems. A wireless system offers communication advantages between a single end and a large area with less effort and flexibility, whereas a wired system is well-suited only for end-to-end communication. Furthermore, polarization and incidence angle insensitivity characteristics of the MTMs result in their wide application for energy conversion in energy harvesting systems. MTM-inspired advancements are emerging as a trivial area in wireless scenarios. MTMs are materials having novel or engineered configurations that create electrical and magnetic characteristics that are unusual or difficult to get in nature. Because of their unique electromagnetic properties, such as group velocity and phase variations, MTMs have sparked widespread interest due to their potential for engineerable permittivity (ε), permeability (μ), and negative refractive index, which has led to their potential use in a variety of electromagnetic scenarios spanning from microwave to optical, especially in radiated-wave equipment, such as reader fabrications [1].

Before going any further, it is necessary to briefly clarify the meaning and genesis of the term MTM. MTMs are a class of "engineered" media with unique electromagnetic characteristics not found in nature. "Metamaterial" is derived from the Greek term "meta," which means "above" or "beyond." Compared with the electromagnetic characteristics of commonly existing materials, the metamaterials can vary their electromagnetic properties to anything. Metamaterials are manufactured periodic media whose periodicity is a fraction of the input electromagnetic wave's wavelength [2]. The qualities of metamaterial are derived from its structure rather than its content. The probability of the physical existence of a material having negative permittivity, permeability, and the refractive index was put forward in the 1960s, and the credit goes to a Russian physicist, Victor Georgievich

Veselago. Maxwell's equation contributes to a significant part in permitting such media, and the parameters representing the propagation of plane waves inside them are (i) electric field intensity E, (ii) the magnetic field intensity H, and (iii) the wave vector k. In contrast to a right-handed triplet generated by a typical right-handed media (RHM), E, H, and k produce a left-handed triplet.

Therefore, the former media is known as the left-handed media. Both of the configurations are depicted in Figure 1. E, H, and k comprise a left-handed triplet, whereas E, H, and S (Poynting vector) involve a right-handed association [1]. As a result, the k in LHM is anti-parallel to S, energy travels outwards from the source, and the phase is advanced.

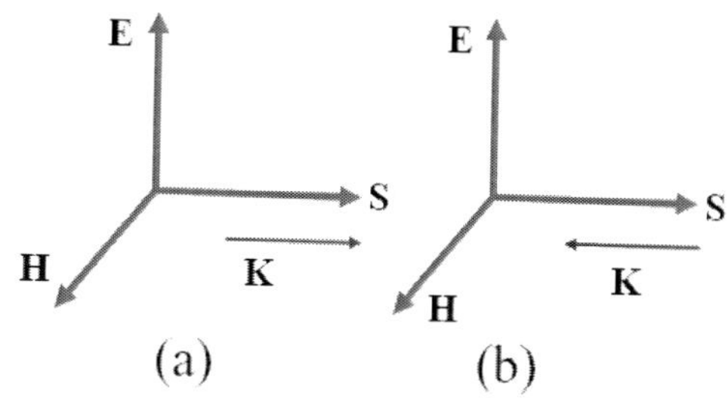

Figure 1. Right and left-handed media Coordinates.

2. Metamaterial Inspired RFID

RFID technology uses low-power radio transmissions to collect data from a remote device. An RFID system comprises two main parts: an RFID reader and a transponder. The interrogator transmits a signal to the transponder, which is reflected or backscattered by the transponder. The middleware interprets and sends the data to the host computer in the appropriate format. It can also be transported to any location on the planet over the internet. The working principle of an RFID system is depicted in Figure 2. Despite its numerous advantages, such as the lack of a direct line of sight requirement, increased tracking distance, and higher encoding capacity, RFID has yet to replace barcodes due to its cost. In this context, the chipless tag [3], which lacks a chip, is a cost-effective alternative to the barcode. An antenna receives the signal from the RF source at the radio frequencies and is further converted voltage/current for information tracking and further processing. The factors which influence the energy conversion efficiency are as follows: (i) mismatch in polarisation, (ii) sensitive nature to the angle of incidence, (iii) misalignment, (iv) impedance imbalance, and (v) nonlinear effects of the rectifying circuit. It is a challenging task to enhance this efficiency. Metamaterials have been created to perform functions that would be difficult to do with conventional natural elements. The multiple resonance feature is ideal for a variety of high-coding capacity RFID applications [4, 5].

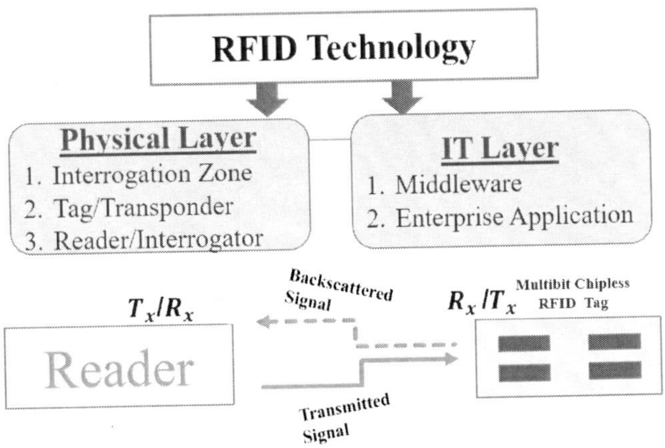

Figure 2. Layered architecture.

The interest in MTMs has grown dramatically over the last decades due to their numerous applications, from low microwave to optical frequencies [6-7]. There are a variety of commercial uses for metamaterials on the market, such as metamaterial-enabled antenna design for space applications. Metamaterials were used in radomes and radar system equipment. They were recently designed for 5G and self-driving vehicle applications. Metasurfaces for magnetic resonance imaging, non-invasive glucose detection, and radio wave imaging are some of the newer applications of metamaterials [8-9]. Aside from the tag design, the tag ID recognition procedure is a hurdle in chipless RFID systems.

Thoughtful knowledge of the behavior through the radar cross-section (RCS) is necessary to design a chipless RFID tag effectively. The RCS of a transponder can be represented as [10]:

$$\sigma = \frac{\lambda^2 R_{ai} G_{tag}^2}{\pi \,|\, Z_{ai} + Z_{ic} \,|} \tag{1}$$

where λ denotes the operating signal's wavelength. R_{ai} is the real part of antenna input impedance Z_{ai}, Z_{ic} denotes chip impedance for a chipped RFID tag and G_{tag} is the tag antenna gain. As a chipless RFID has no chip, Z_{ic} can be assumed to be 50 Ω and Z_{ai} is real. Now RCS of the chipless RFID tag [10] at the resonant frequency can be reduced to (2):

$$\sigma_{ctag} = \frac{\lambda^2 R_{ai} G_{tag}^2}{\pi |Z_{ai} + Z_{ic}|} = \frac{\lambda^2 G_{tag}^2}{2\pi} \tag{2}$$

The maximum reading distance for a chipped tag is defined by [10]:

$$RR_{\max} = \frac{\lambda}{4\pi} \sqrt{\frac{G_{reader} PLF_{tag} G_{tag,normalized}}{P_{i,\min}}} \tag{3}$$

where *RR* denotes the read range, *G* represents the gain, and the suffix have their usual meaning. *PLF* is the polarization loss factor [10].

A split-ring resonator (SRR) is a metamaterial structure created intentionally. Their goal is to achieve the necessary magnetic susceptibility (magnetic response) in various forms of microwave-frequency metamaterials. A periodic array of split ring resonators produces the requisite strong magnetic coupling to an applied electromagnetic field, which is not readily obtainable in traditional materials, for e.g., negative permeability. The electromagnetic field has a distinct effect on each SRR. The loops can be concentric or square, and the gaps between them can be adjusted as desired. The equivalent circuit of an SRR is shown in Figure 3 [11-12]. A coupling capacitor of capacitance, *C*, and a transformer having a transforming ratio of *n* are utilized to model the capacitive and inductive coupling. The circuit model of the single ring configuration is illustrated in Figure 3. The circuit is formed utilizing a simple RLC resonator having a resonant frequency of

$$\omega_0 = \frac{1}{\sqrt{LC}} \tag{4}$$

where, ω_0 represents the resonant frequency, *L* and *C* are the mutual coupling inductance and capacitance. Because the dimensions of the two rings are very close to each other, $L1 \approx L2 \approx L$ and $C1 \approx C2 \approx C$, the double SRR (split-ring resonator) is approximately identical to the single SRR provided the mutual coupling is low. This leads to an overall resonant frequency approximately equal to that produced by a single SRR of identical size. However, a high current density is a reason for generating a high magnetic moment.

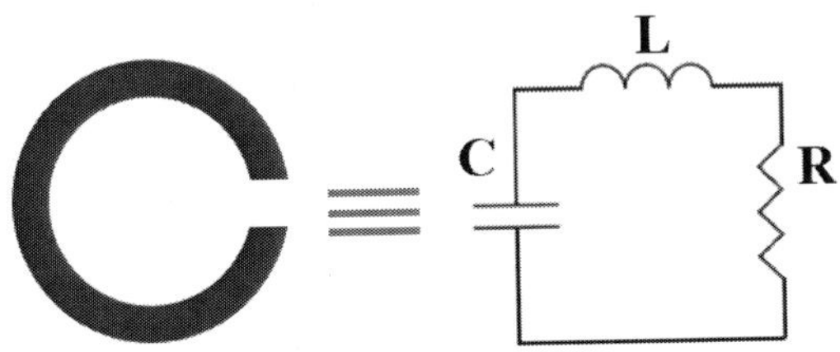

Figure 3. Equivalent Circuit of MTM inspired SRR.

3. Metamaterial Inspired High-Capacity Multi-State Resonators

Chipless RFID tags can benefit from a compact, high-coding-capacity MTM-based resonator. This type of tag consists of various resonating elements and cross-polarized ultra-wideband antennas. The multiple MTM-inspired resonators comprise a U-shaped connected microstrip structure having separated (*K*-1) legs, where *K* is the number of resonating structures used to design the transponder/tag. One of the legs is intended to be attached to the U-frame through a metallic strip to alter the resonance frequency. This function allows you to change the frequency of each resonator to any *K*-resonance frequencies.

As a result, each resonance represents one of the resonator's *K* states. This enables each resonator to represent several bits of data. The tags consisting of *N* resonators have a coding capacity of *KN*. The structure presents the benefit of high information storing capability with a compact dimension, and the detection procedure is effortless by capturing *N* resonant frequencies for every code. In contrast to the conventional one-bit resonator tags (two states) having a coding capacity of 2*N* with *N* resonators, the presented structure has an enhanced data encoding capability. Moreover, to improve spectral efficiency, a different configuration is described that permits the bandwidth of every single resonator to be shared across two resonators.

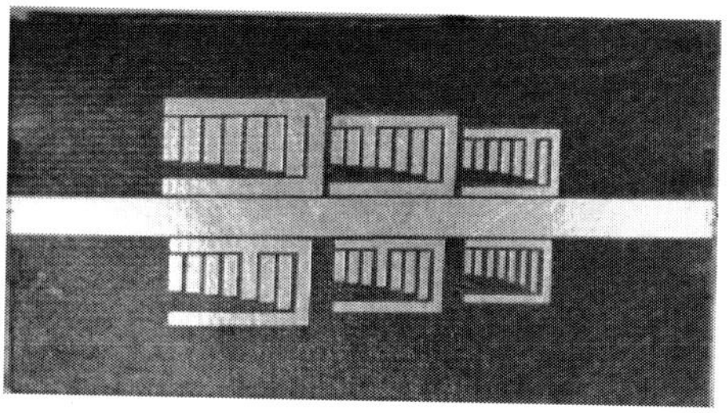

Figure 4. Fabricated MTM-based high capacity tag (copyright [13]).

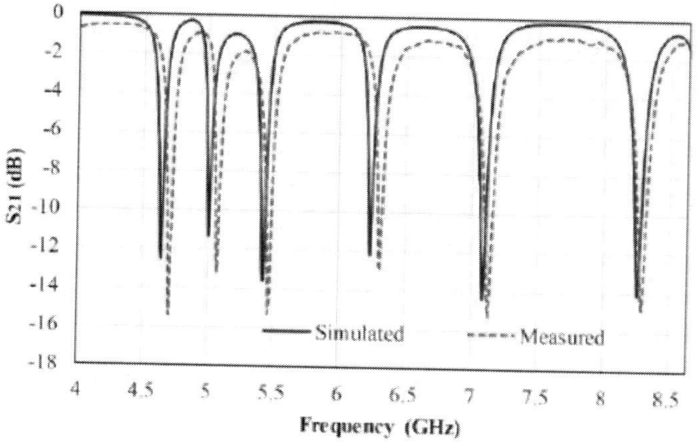

Figure 5. Simulated and Measured S_{21} (copyright [13]).

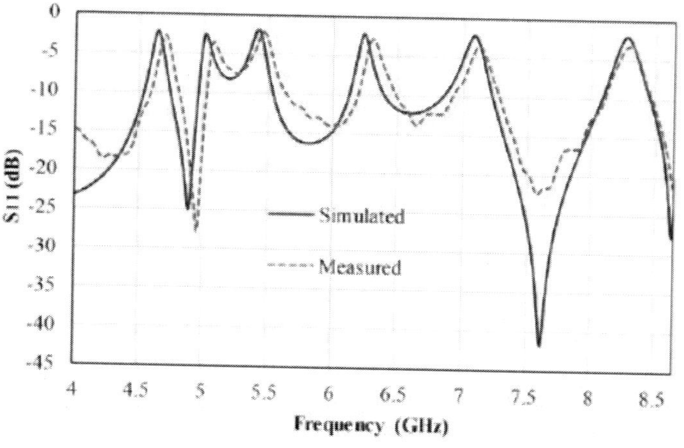

Figure 6. Simulated and observed S_{21} and S_{11} responses (copyright [13]).

Many tags are devised and implemented on the RT Duroid 5880 substrate as proof of concept. An 8-state version of the suggested resonator may be built in a space that is roughly the same size as a typical 5-GHz 2-state resonator (fabricated prototype shown in Figure 4). The empirical and simulated findings are then in good accord [13].

The resonating elements are placed accordingly to prevent the loading effect, which, if present, results in a frequency shift when more than one arm is coupled to the linked line. The simulated and measured insertion loss (S_{21}) of the fabricated prototype is displayed in Figure 5, and the return loss (S_{11}) of the manufactured configuration is shown in Figure 6. 4.42 GHz, 4.66 GHz, 5.30 GHz, 5.67 GHz, 6.41 GHz, and 6.91 GHz were the recorded resonance frequencies for this code. The second technique employs varying arm lengths to reduce the difference between nearby resonance frequencies to around 70 MHz rather than 100 MHz. Two 6-resonator circuits are designed and implemented, each with a different code. Figure 6 summarises the simulated and measured S_{21} and S_{11} performances for these two circuits. The resonance frequencies are well distinct and the span of the frequency range is from 4.7 to 8.3 GHz.

In this condition, an enhancement of the spectral density is observed between 1 bit/290 MHz and 1 bit/200 MHz. An estimated spectral density greater than 1 bit/120 MHz is obtained by combining the approaches described above [13]. A chipless RFID tag with a small reconfigurable MTM–based K-state resonator facilitates the representation of multiple bits by every resonator. The structure offers the advantage of surging the tag's capability without increasing the number of resonating elements along with a simple detection procedure. As a result, the suggested design enables the storage of a large amount of data in a little space, lowering the cost of storage per bit. In [14] authors present a chipless RFID reader antenna based on MTM and a modified rectangular resonator. The RFID reader antenna is fabricated and measured to validate multiple narrowband resonance characteristics. It is observed that the antenna can capture data from a multi-frequency coded chipless tag with a data capacity of 11 bits. The chipless tag has gains of 4.7 dBi at 0.58 GHz and 5.2 dBi at 2.4 GHz. In addition, [15] examines two passive transponders based on the MTM-inspired resonator concept, with an encoding capability of three bits. Comparing the RCS and frequency coding approaches reveals the decrease of tag dimensions during manufacturing. Multi-bit encoding capability is crucial for the efficient identification and tracking that metamaterial-based resonators can achieve.

Furthermore, this advancement contributes to the efficient design of compact passive chipless RFID transponders. The maximum RCS for transponders 1 and 2 were close to -23 dBm. It lowers the cost of tag production, making it suitable for mass industrial production [15].

4. MTMS-Based Compact Flexible Chipless Tag

The MTM inspired Modified Complementary Split Ring Resonator (MCSRR) is a successful technology for producing the miniaturized flexible chipless tag having an enhanced coding capability that encodes information based on the frequency signature. In [16], five slotted overlapping MCSRR with different widths and a diameter of 48 mm × 48 mm are presented as a 19-bit chipless transponder where a frequency shifting approach is employed. A flexible (Polyethylene Terephthalate) PET substrate with a permittivity of 0.2 was used to create the structure. The operational frequency ranges from 0.9 to 2.7 GHz.

The omnidirectional tag antenna is integrated into the proposed MCSRR framework using the retransmission measurement method. A log-periodic antenna with 7 dBi gain is used to enhance the tracking distance from the interrogator to the transponder. The chipless tag that consists of five MCSRR components can encode 19 bits. A reading distance of 30 cm with 30 dBm transmission power is essential for identifying the above-mentioned chipless tag [16]. High Q-factor and narrow band resonators significantly enhance the bit capacity of these compact chipless tags [17]. The following relation [17] can be used to express the real size and BW.

$$\text{Actual size per bit}\left(\frac{\lambda^2}{bit}\right)=\frac{\text{Dimension}\left(\lambda^2\right)\text{ based on mid frequency}}{\text{Number of bits}} \tag{5}$$

$$\text{BW allocation per bit}\left(\frac{GHz}{bit}\right)=\frac{\text{BW (GHz)}}{\text{Number of bits}} \tag{6}$$

A metamaterial is an engineered substance with extraordinary electromagnetic properties that aren't present in nature. Artificial materials having the property of switching between positive and negative values of permittivity and permeability are well-suited to meet the requirement in the advanced microwave equipment like antennas [18], filters [19], lenses [20], and passive sensors [21]. This will empower the researchers to make smaller designs with multiband behavior and better device performance. The SRR and CSRR's unique metamaterial features have been used in microwave design, such as antennas, filters, absorbers, and couplers, to achieve small downsizing [22] and multiband behavior [23], and good performance [24]. The gap created between the inner and outer rings of the resonator greatly influences the resonant frequency. Simultaneously, the initial resonant frequency has been lowered due to increased distributed capacitance and inductance between both rings. This criterion is required for size reduction.

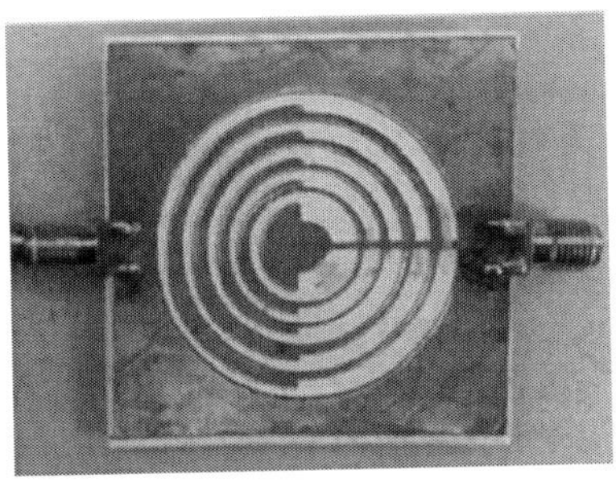

Figure 7. Fabricated Prototype (MTM inspired) (copyright [16]).

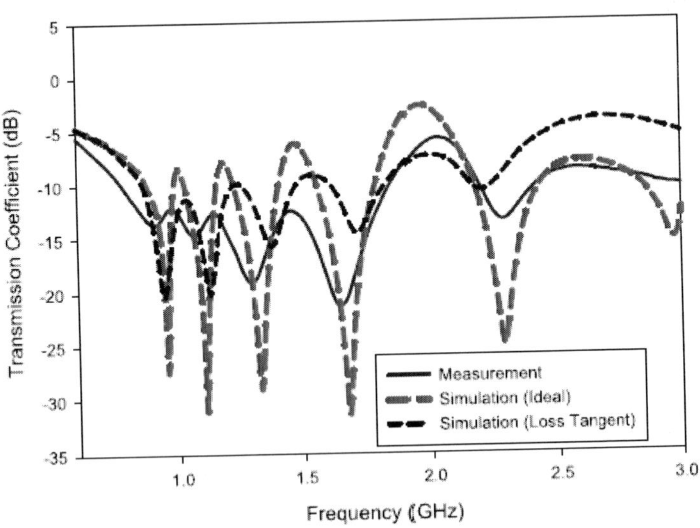

Figure 8. Transmission coefficient (simulated and measured) (copyright [16]).

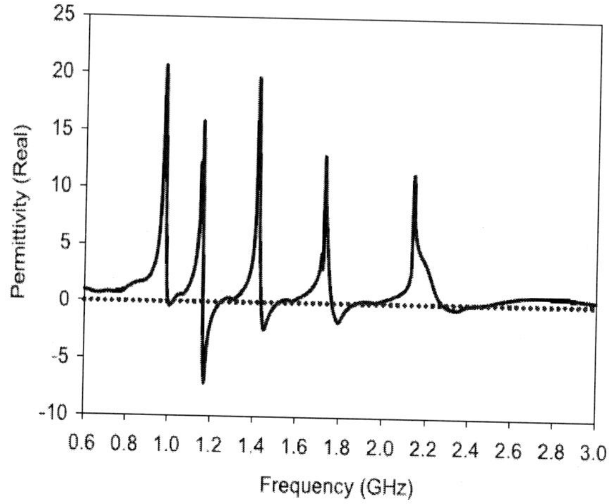

Figure 9. Effective permittivity (copyright [21]).

Figure 7 shows the fabricated layout of a flexible, small, and high-capacity passive transponder on PET substrate with varied width architectures (based on MCSRR). On the top layer, a 50 Ω transmission line of width 1.3mm was placed, and in the middle ground, every resonating element is a ring resonator with a gap. The overlying slotted MCSRR promotes dimension reduction by using the ground plane's total space and lowering the operating frequency by minimizing the separation frequency of the two resonators. Figure 7 design represents the five modified complementary split-ring resonators with different widths. Each ring resonator's split gap will be modified based on the chipless RFID tag's particular coding value. From Figure 8, it can be observed that the simulation and measurement of S_{21} results are closely associated. The measured resonant frequency marginally moves to 0.91 GHz, 1.07 GHz, 1.31 GHz, 1.66 GHz, and 2.28 GHz, while the computed ideal resonant frequency works well at 0.94 GHz, 1.10 GHz, 1.33 GHz, 1.68 GHz, and 2.29 GHz.

A substrate with a high loss tangent of about 0.35 will result in the fall of the level of the transmission coefficient of the measured resonant frequency. Using the robust technique [25] from the simulated S-parameter, the effective permittivity is obtained based on the design. The actual permittivity curves give the resonance qualities. As demonstrated in Figure 9, negative permittivity will exist in frequency ranges for CSRR metamaterial structure, which is advantageous for RFID operation with flexibility, large data capacity, and downsized transponder with a low frequency of operation.

5. Challenges and Research Directions

The reader in a passive chipless RFID system must handle the challenges listed below to vie with optical barcodes and chipped RFID. Each challenge is presented in order.

5.1. Size and Price

Although chipless tags are less expensive than chipped tags, the reader is more expensive due to the antenna elements that operate in the UWB spectral region. The introduction of metamaterial technology is one way to minimize the cost as well as the footprint of the reader. A single-antenna topology can also be used instead of the two-antenna reader. Another strategy is using simpler architectures with fewer broadband microwave modules [26]. The trade-off between the cost and the bandwidth of the reader is significant. A good reader appealing to business is the one with excellent performance and cost-efficient. So, a low-cost reader that is operable in the entire UWB range of frequencies is in high demand. Once a new platform for chipless RFID is implemented, it considerably impacts the reader's pricing [27].

5.2. Reading Speed and Accuracy

Detection time, especially when a fast-moving object can be identified, is a crucial element in some applications. In a frequency-based approach, the detection time for passive RFID scanners remains an unsolved issue. The reading time and identification accuracy in frequency domain readers are mostly determined by the hardware design, operational frequency, and resolution [28]. Furthermore, the narrow spacing between the resonance frequencies in high-capacity tags affects the reading accuracy. The chipless tag can be adjusted to improve detection performance in this situation. Transponders with high-quality factors produced on a low-loss material are used to minimize the possibility of detection mistakes [27] [29] at a lower cost. Instead of using expensive substances for generating tags, improving the reader's hardware and software is more practical to improve reading accuracy. For this reason, a signal processing algorithm is advised to be developed in the reader's system.

5.3. Sensitivity to the Tag Alignment

The polarisation inclination of the investigation signals determines the tag's behavior in most configurations. The system's performance depends on the chipless tag's orientation; to get the backscattered signal from the tag in different orientations, the reader architecture must be changed. In this case, the use of a reader with a circular–polarized antenna or two dual-polarized antennas could be a better preference [30-31].

5.4. Reading Distance and Interferences

If the range can be extended, the chipless RFID technology can be utilized to replace the chipped equivalent for various long-range applications. However, various limitations limit the system's reading range. Wideband noisy spectrum, quasi-chipless tags, and UWB regulatory limits are only a few examples. The electromagnetic signal leakage from the interrogators, transponders and the reflected signal from other sources are sources of interference signals, and it is not easy to eliminate them.

This is because in small-reading chipless RFID systems, the broadcasted and reflected waves are said to interact in both time and frequency domains, making it difficult to discriminate between them. Consequently, boosting the system's detection limit and adjusting for jamming signals on the reader side can enhance the effective detection area. To address the problem, a UWB adjustment module [32] for eliminating steady sources of interference and signal processing methods for neutralising dynamic interferences [33] could be used.

5.5. Anticollision

One of the most challenging tasks for RFID devices in real applications is reading many tags close to the reader. This results in a collision among the tags to reply to the investigating signal at a time. Hence, other methods are incorporated to avoid interference of other tags' responses. These strategies are primarily used in the reader's middleware [34]. But the available documented strategies for anticollision were not used on more but fewer tags, and the equipment needed to work with is expensive. For instance, in [35] an algorithm that works in frequency signature-based approach of 3-

transponders, for exact signature detection vector network analyzer is used. Moreover, the method mentioned above uses background subtraction methodology, which cannot be a good fit for practical uses, especially moving tags. As a result, this field of study is yet to be investigated. It should be observed that several of these techniques require a hardware architecture, which adds to the reader's complexity.

5.6. Compatibility with Regulations

The reader's optimum EIRP should be acceptable to the UWB protocol specification. Because of the short interrogation pulse (a few nanoseconds), this requirement does not affect the performance of time-domain readers but limits the effective reading range of frequency-domain readers. Several techniques are found in the literature for increasing the EIRP level while keeping compliant with UWB rules [3, 36].

5.7. Improved Data Capacity for MTM Tags

The range of frequencies over which the chipless tags mainly operate under the UWB range imposes a limitation on the total number of encoded bits. To improve the bit capacity MTM based tags can be a suitable option [37]. The development of metamaterial-inspired tags that can give high data storage capacity is vital in research. The additional hardware requirement is the factor that needs to be considered to set the price and fabrication difficulty of the interrogator design in order to increase the coding capacity of the transponder and to balance between the cost and the data handling capacity. MTM inspired hybrid tags must be designed with great care to ensure that the reader's cost and complexity are kept to a minimum.

Conclusion

The application of the influential properties of the metamaterials in chipless RFID is regarded as a milestone in developing the features like coding capacity, compact dimension, and RCS. Different kinds of metamaterial-inspired resonators are discussed to reduce the cost and the physical dimension of the communication devices. Different design methodologies, along with their critical analysis, are presented in this chapter for real field applications. MTM-based structures are highly beneficial for RFID and many microwave applications in the retail and healthcare industry, improving the spectral bit capacity by limiting the number of resonators. The challenging aspects in the implementation of the technology and the further research direction are elaborated from the reader's point of view.

References

[1] Veselago, Victor Georgievich. "The electrodynamics of substances with simultaneously negative values of ε and μ." *Physics-Uspekhi* 10, no. 4 (1968): 509-514.

[2] Holloway, Christopher L., Edward F. Kuester, Joshua A. Gordon, John O'Hara, Jim Booth, and David R. Smith. "An overview of the theory and applications of metasurfaces: The two-dimensional equivalents of metamaterials." *IEEE antennas and propagation magazine* 54, no. 2 (2012): 10-35.

[3] Finkenzeller, Klaus. *RFID handbook: fundamentals and applications in contactless smart cards, radio frequency identification and near-field communication*. John wiley & sons, 2010.

[4] Pendry, John Brian. "Negative refraction makes a perfect lens." *Physical review letters* 85, no. 18 (2000): 3966.

[5] Smith, David R., Willie J. Padilla, D. C. Vier, Syrus C. Nemat-Nasser, and Seldon Schultz. "Composite medium with simultaneously negative permeability and permittivity." *Physical review letters* 84, no. 18 (2000): 4184.

[6] Sievenpiper, Dan, Lijun Zhang, Romulo FJ Broas, Nicholas G. Alexopolous, and Eli Yablonovitch. "High-impedance electromagnetic surfaces with a forbidden frequency band." *IEEE Transactions on Microwave Theory and techniques* 47, no. 11 (1999): 2059-2074.

[7] Binion, J. Daniel, Erik Lier, Thomas H. Hand, Zhi Hao Jiang, and Douglas H. Werner. "A metamaterial-enabled design enhancing decades-old short backfire antenna technology for space applications." *Nature communications* 10, no. 1 (2019): 1-7.

[8] Sihvola, Ari. "Metamaterials in electromagnetics." *Metamaterials* 1, no. 1 (2007): 2-11.

[9] Smith, David R., John B. Pendry, and Mike CK Wiltshire. "Metamaterials and negative refractive index." *Science* 305, no. 5685 (2004): 788-792.

[10] Karmakar, Nemai Chandra, ed. *Handbook of smart antennas for RFID systems*. John Wiley & Sons, 2011.

[11] Pendry, John B., Anthony J. Holden, David J. Robbins, and W. J. Stewart. "Magnetism from conductors and enhanced nonlinear phenomena." *IEEE transactions on microwave theory and techniques* 47, no. 11 (1999): 2075-2084.

[12] Costa, Filippo, Agostino Monorchio, and Giuliano Manara. "Efficient analysis of frequency-selective surfaces by a simple equivalent-circuit model." *IEEE Antennas and Propagation Magazine* 54, no. 4 (2012): 35-48.

[13] Abdulkawi, Wazie M., and Abdel-Fattah A. Sheta. "K-state resonators for high-coding-capacity chipless RFID applications." *IEEE Access* 7 (2019): 185868-185878.

[14] Mishra, Durga Prasad, and Santanu Kumar Behera. "Modified rectangular resonators based multi-frequency narrow-band RFID reader antenna." *Microwave and Optical Technology Letters* 64, no. 3 (2022): 544-551. https://doi.org/10.1002/mop.33116.

[15] Mishra, Durga Prasad, and Santanu Kumar Behera. "A Novel Technique for Dimensional Space Reduction in Passive RFID Transponders." *2021 2nd International Conference on Range Technology (ICORT)*, pp. 1-4. IEEE, 2021.

[16] Jalil, Mohd Ezwan B., Mohamad Kamal A. Rahim, Himdi Mohamed, Noor Asmawati Binti Samsuri, Noor Asniza Murad, Raimi Dewan, Huda Bin A. Majid, Nur Biha Mohamed Nafis, Levy Olivia Nur, and Bambang Setia Nugroho. "High capacity and miniaturized flexible chipless RFID tag using modified complementary split ring resonator." *IEEE Access* 9 (2021): 33929-33943.

[17] Wang, Lu, Ting Liu, Johan Sidén, and Gang Wang. "Design of chipless RFID tag by using miniaturized open-loop resonators." *IEEE Transactions on Antennas and Propagation* 66, no. 2 (2017): 618-626.

[18] Zhang, Jiahao, Sen Yan, and Guy AE Vandenbosch. "Metamaterial-inspired dual-band frequency-reconfigurable antenna with pattern diversity." *Electronics Letters* 55, no. 10 (2019): 573-574.

[19] Tan, Cheng, Yu Wang, Zhongming Yan, Xinyi Nie, Yonghai He, and Weirong Chen. "Superconducting filter based on split-ring resonator structures." *IEEE Transactions on Applied Superconductivity* 29, no. 4 (2019): 1-4.

[20] Orazbayev, B., V. Pacheco-Pena, M. Beruete, and Miguel Navarro-Cía. "A self-supporting broadband zoned fishnet metamaterial lens operating at the millimeter-wave V-band." *2015 9th International Congress on Advanced Electromagnetic Materials in Microwaves and Optics (METAMATERIALS)*, pp. 241-243. IEEE, 2015.

[21] Dong, Lei, Li-Feng Wang, and Qing-An Huang. "Enhancing the readout of passive wireless sensors by using left-handed metamaterials." *2016 IEEE SENSORS*, pp. 1-3. IEEE, 2017.

[22] Ramzan, Mehrab, and Kagan Topalli. "A miniaturized patch antenna by using a CSRR loading plane." *International Journal of Antennas and Propagation* 2015, no. 2015 (2015): 1-9.

[23] Paul, Princy Maria, Krishnamoorthy Kandasamy, and Mohammad S. Sharawi. "A triband circularly polarized strip and SRR-loaded slot antenna." *IEEE Transactions on Antennas and Propagation* 66, no. 10 (2018): 5569-5573.

[24] Peng, Liang, Shuaifei Sang, Zhuoyuan Wang, Huayan Jin, Aiting Wu, Kuiwen Xu, and Gaofeng Wang. "Wideband radiation from an offset-fed split ring resonator with multi-order resonances." *IEEE Antennas and Wireless Propagation Letters* 17, no. 12 (2018): 2198-2202.

[25] Chen, Xudong, Tomasz M. Grzegorczyk, Bae-Ian Wu, Joe Pacheco Jr, and Jin Au Kong. "Robust method to retrieve the constitutive effective parameters of metamaterials." *Physical review E* 70, no. 1 (2004): 016608.

[26] Forouzandeh, Mohammadali Mazyar, and Nemai Karmakar. "Towards the improvement of frequency-domain chipless RFID readers." *2018 IEEE Wireless Power Transfer Conference (WPTC)*, pp. 1-4. IEEE, 2018.

[27] Aliasgari, Javad, Mohammadali Forouzandeh, and Nemai Karmakar. "Chipless RFID readers for frequency-coded tags: Time-domain or frequency-domain?." *IEEE Journal of Radio Frequency Identification* 4, no. 2 (2020): 146-158.

[28] Garbati, Marco, Etienne Perret, and Romain Siragusa. *Chipless RFID reader design for ultra-wideband technology: design, realization and characterization*. Elsevier, 2018.

[29] Daliri, Ali, Amir Galehdar, Wayne ST Rowe, Sabu John, Chun H. Wang, and Kamran Ghorbani. "Quality factor effect on the wireless range of microstrip patch antenna strain sensors." *Sensors* 14, no. 1 (2014): 595-605.

[30] Garbati, Marco, Etienne Perret, Romain Siragusa, and Christophe Halopé. "Toward chipless RFID reading systems independent of tag orientation." *IEEE Microwave and Wireless Components Letters* 27, no. 12 (2017): 1158-1160.

[31] Genovesi, Simone, Filippo Costa, Francesco Alessio Dicandia, Michele Borgese, and Giuliano Manara. "Orientation-insensitive and normalization-free reading chipless RFID system based on circular polarization interrogation." *IEEE Transactions on Antennas and Propagation* 68, no. 3 (2019): 2370-2378.

[32] Forouzandeh, Mohammadali, and Nemai Karmakar. "Application of wideband differential phase shifters with wide phase range in chipless RFID readers." *IEEE Transactions on Microwave Theory and Techniques* 67, no. 9 (2019): 3636-3650.

[33] Ramos, Angel, Etienne Perret, Olivier Rance, Smail Tedjini, Antonio Lázaro, and David Girbau. "Temporal separation detection for chipless depolarizing frequency-coded RFID." *IEEE Transactions on microwave theory and techniques* 64, no. 7 (2016): 2326-2337.

[34] El-Hadidy, Mohamed, Ahmed El-Awamry, Abdelfattah Fawky, Maher Khaliel, and Thomas Kaiser. "Real-world testbed for multi-tag UWB chipless RFID system based on a novel collision avoidance MAC protocol." *Transactions on Emerging Telecommunications Technologies* 27, no. 12 (2016): 1707-1714.

[35] Rezaiesarlak, Reza, and Majid Manteghi. "A space–time–frequency anticollision algorithm for identifying chipless RFID tags." *IEEE Transactions on Antennas and Propagation* 62, no. 3 (2013): 1425-1432.

[36] Forouzandeh, Mohammadali Mazyar, and Nemai Karmakar. "Towards the improvement of frequency-domain chipless RFID readers." *2018 IEEE Wireless Power Transfer Conference* (WPTC), pp. 1-4. IEEE, 2018.

[37] Mishra, Durga Prasad, Tanmaya Kumar Das, Santanu Kumar Behera, and Nemai Chandra Karmakar. "Modified rectangular resonator based 15-bit chipless radio frequency identification transponder for healthcare and retail applications." *International Journal of RF and Microwave Computer-Aided Engineering* 32, no. 6 (2022): e23127.

Chapter 3

Metamaterial Absorbers: Theory and Applications

Zian Li, Hui Xia*, Wentao Lei and Yishan Zhou
College of Physics and Electronics, Central South University, Changsha, Hunan, China

Abstract

The chapter focuses on the related theory and application of the metamaterial absorber. The idea of metamaterials is presented, and its benefits are underlined by comparison with the properties of conventional materials. Then the metamaterial absorber is introduced. The basic principle, development process, and application scenario of the metamaterial absorber are also described. Three common methods for analyzing the properties of metamaterials are discussed in detail: impedance matching theory, equivalent circuit theory, and multiple reflection interference theory. Finally, future research prospectives of metamaterial absorbers are introduced for realizing low-frequency ultra-wideband absorption, the multifunctional integrated metamaterials, and the design of three-dimensional metamaterials.

Keywords: metamaterial absorber, impedance matching, reflection interefence theory, equivalent circuit

1. Introduction

Using electromagnetic wave absorbing materials effectively reduces electromagnetic interference because it can convert electromagnetic waves

* Corresponding Author's Email: xhui73@csu.edu.cn.

In: Metamaterials for Microwave and Terahertz Applications
Editors: B. Appasani, O. Prakash Acharya, A. Vidyakant Jha et al.
ISBN: 979-8-88697-330-3

into heat or other forms of energy. The absorbers used in traditional electromagnetic wave absorbing materials mainly include soft magnetic materials (such as spinel ferrite powder) and dielectric materials (such as carbon powder-loaded polymers). However, these electromagnetic wave-absorbing materials have low absorptance, large size, and poor stability. Therefore, according to these defects of traditional electromagnetic wave absorbing materials, designing a new electromagnetic wave absorbing material with light weight, small volume, and high absorptance is particularly meaningful. Based on the traditional electromagnetic wave absorbing materials, the preparation methods were improved, and new materials such as nanomaterials [1], chiral materials [2], electrochromic materials [3], conductive polymers [4], and electromagnetic metamaterials [5] were obtained. Electromagnetic metamaterials have become a research hotspot because of their anomalous and tunable properties. Electromagnetic metamaterials can maintain the high absorptance of electromagnetic waves and provide a novel idea for developing absorbing materials because of their adjustable dielectric constant and permeability.

A metamaterial is an artificially designed structure or material. Just as matter is composed of atoms, the structural elements of metamaterial are also made of traditional materials of ordinary atoms. Its composition is no different from that of ordinary materials. The main difference is that its material type, quantity, size, and layout can be customized. The structural elements constituting metamaterials can be artificially designed into any size and shape, such as split-ring resonator, coupled split-ring resonator, bending structure, etc [6]. Metamaterials can break the shackles of natural laws by orderly designing the structure on the key scale of materials and have some abnormal physical properties. For instance, negative refractive index [7], abnormal doppler effect [8], and super lens [9]. These anomalous physical properties of metamaterials enable to control and manipulation of the transmission of longitudinal waves (such as light, electromagnetic and sound waves) to some extent. The discovery of metamaterials has opened up a new research field for developing electromagnetic theory. It has great scientific significance and broad application prospects for economic promotion and social progress. Therefore, metamaterials have become one of the research hotspots in international academic circles for more than ten years. In the early research stage, studies on metamaterials almost concentrated on the GHz band due to the limitations of experimental conditions. However, with the development of nanotechnology, the research on the absorber is no longer satisfied with the GHz band but develops to the higher frequency THz band. At the same time, metamaterials are also used in the improvement of microwave equipment and optical equipment, among which the outstanding achievements are microwave absorbers, filters, detectors and optical switches.

1.1. Applications of Metamaterials

Based on the Maxwell equation, the propagation of electromagnetic waves in materials with negative dielectric constant and permeability were studied [10], and the electromagnetic characteristics were discussed: the direction of phase velocity and group velocity is opposite, electric field intensity (E), magnetic field intensity (H) and wave vector (k) are expressed as left-handed spiral rule, which is just opposite to the right-handed spiral rule when electromagnetic waves propagate in traditional materials. Therefore, this material is defined as left-handed material. However, at that time, the material could not be observed in nature and could not be used. Subsequently, Pendry theoretically reported the low-frequency plasma periodic array with a negative dielectric constant. Smith and Schultz conducted an experimental study on negative refractive index materials. The materials were composed of periodic cells with negative dielectric constant and negative permeability. These periodic cells showed left-hand characteristics in the form of metal short lines and open rings. In recent years, metamaterials have become popular because of their versatility in regulating incident electromagnetic waves and some characteristics that do not exist in nature. For instance, negative refractive index [11], extraordinary optical transmission characteristics [12], electromagnetically induced transparency [13, 14] and electromagnetic stealth cloak [15]. According to its characteristics, it has many applications. Zhou et al. studied all-dielectric metamaterials based on Mie scattering theory and proposed tunable metamaterials based on liquid crystals [16]. Zhao et al. studied the design and preparation of tunable metamaterials based on electrorheological fluids [17]. Zhang et al. conducted research on chiral metamaterials and broadband negative refractive index metamaterials based on transmission line theory [18]. Zhou et al. studied surface plasmon metamaterials based on the effects of polarization rotation, directional radiation and sub-wavelength imaging of metamaterials [19]. Ran et al. designed metamaterials based on S-shaped, rectangular and other metal structures, and studied the design and optimization of metamaterial stealth cover in theory [20]. Cui et al. mainly focused on the research and preparation of new microwave equipment, such as carpet covers, antenna plane lens, and omnidirectional electromagnetic wave absorption structures [21].

2. Metamaterial Absorber

As mentioned above, metamaterials have made significant achievements in photonics due to some unnatural electrodynamic properties and effects [22-26]. With the development of nanotechnology, many applications have been derived from metamaterials, one of the outstanding achievements in electromagnetic metamaterial absorbers [27]. It can be used to improve the efficiency of absorbing solar energy [28], can be applied to plasma sensors [29], radiant heat meters [30], wireless energy transmission [31], and a perfect light absorber has been gradually designed [32-34].

The only way to realize the electromagnetic wave's perfect absorption is to completely eliminate the reflection on the interface. However, it is difficult to match free space's impedance with the material's input impedance. Traditional electromagnetic absorbing materials usually use multiple reflections and interferences to simultaneously eliminate the reflected waves at the interface to improve the absorptance. For example, as can be seen from Figure 1-1 (a), the Salisbury screen composes of a very thin top resistive screen, a middle dielectric layer with a thickness of 1/4 wavelength, and a bottom metallic plane [35]. When the electromagnetic wave is vertically incident to the Salisbury screen, the electromagnetic wave reflected by the top surface resistance and the electromagnetic wave reflected by the bottom metal plate has the same size and opposite phase, which makes the two reflected electromagnetic waves produce interference cancellation effect. Therefore, the incident electromagnetic wave is completely bound in the absorber, and part of it is dissipated in the surface resistance so that the absorptance of the electromagnetic wave reaches the maximum. However, Salisbury screens have significant disadvantages: large volume and narrow bandwidth, limiting their application range. Because the bandwidth of the single-layer structure is too small, it cannot meet the demand gradually, the Jaumann absorber of multi-layer structure is designed. This absorber combines a resistance layer in the Salisbury screen and a dielectric loss layer with a frequency selective surface structure. However, this structure still requires a dielectric layer of about 1/4 wavelength thickness, which makes the thickness and weight of the structure huge. For the traditional electromagnetic absorber, the reflection of the low impedance ground plane produces a π phase shift, and the dielectric layer with a thickness of 1/4 wavelength will also provide an additional π propagation phase. Therefore, the total phase changes to 2 π (or 0) during the round-trip propagation, and the phase-matching condition is satisfied in the process of multiple reflection interference. If a metamaterial surface is added to the grounding plate separated by a very thin dielectric layer, a low impedance grounding plate (electrical conductor) can be transformed into a high impedance grounding plate (magnetic conductor). As designed by Engheta [36], the thickness of the absorber is significantly reduced compared with before. If the loss of metal and dielectric materials is considered and the metamaterial structure with appropriate thickness dielectric layer is used, the metamaterial structure shown in Figure 1 (b) can be used as an ideal absorber.

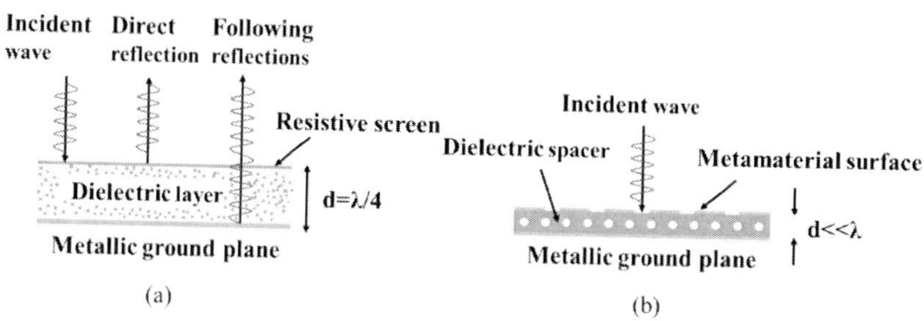

Figure 1.1. (a) Salisbury screen, (b) perfect metamaterial absorber.

Landy proposed the perfect metamaterial absorber for the first time. Compared with the previous design, it cleverly solves the problems of excessive volume and thickness [37]. Then, many perfect metamaterial absorbers were used for distinct application fields. For instance, thermal imaging [38, 39], solar cell [40], plasma sensor [41], radiant calorimeter [30], wireless energy transmission [42] and antenna [43]. A typical absorbing metamaterial is a subatomic material that causes energy consumption. The capacitance and inductance in the equivalent circuit corresponding to the absorber structure are used as energy storage elements. The energy is stored in them when resonance occurs and then dissipates energy through ohmic loss and dielectric loss so as to realize the perfect absorption of electromagnetic waves. The experimental results show that the energy is mainly consumed in the dielectric loss. Therefore, the magnetic resonance with the reverse parallel current has been developed to produce significant dielectric loss [44, 45]. Recently, the perfect application of metamaterial absorber has been further expanded in the frequency range from radio frequency [31, 46] to visible light

[47]. For instance, multi-band absorption [48-50], broadband absorption [51-53], polarization-insensitive [54-56], wide incident angle [57, 58], controllable band [59, 60] and absorption tunable [61, 62]. Generally speaking, the sandwich structure of the metamaterial absorber is the most common structure. The structure consists of three parts: top metal patterns, a middle dielectric layer, and a bottom metal plate. The metal patterns are the main part of the metamaterial absorber; different metal patterns constitute the electromagnetic resonators. The middle dielectric layer always uses silicon dioxide, which mainly plays a supporting role. The bottom metal plate is used to prevent the transmission of the incident wave and form an antiparallel surface current with metal patterns. The absorption can be calculated using the formula:

$$A(\omega) = 1\text{-}R(\omega)\text{-}T(\omega) \qquad (1)$$

where A, R, and T represent the absorptance, reflectivity, and transmissivity of electromagnetic wave, respectively. $R(\omega) = |S_{11}|^2$ and $T(\omega) = |S_{21}|^2$, $|S_{11}|$ and $|S_{21}|$ are the scattering parameters of reflection and transmission respectively.

When the input impedance of the material matches the impedance of the free space ($Z(\omega)=\sqrt{\mu(\omega)/\varepsilon(\omega)}=1$), the perfect absorption is realized, and the general method of impedance matching is to optimize the structural parameters. For maximum absorptance, the reflectivity and transmissivity should tend to 0. Among them, $T = 0$ is easy to achieve by choosing a thick metal plate. This is because if the thickness of the metal plate is much greater than its skin depth, the transmitted wave will be completely localized on the upper surface of the metal plate, which is equivalent to blocking the transmission of wave, so that the transmissivity $T = 0$. Now, only the reflectivity R need to be considered, so $A(\omega) = 1\text{-}R(\omega) = 1\text{-}|S_{11}|^2$ can be used to calculate the absorptance of the electromagnetic wave. The interaction between the incident wave and metamaterial may be caused by multiple reflections [63] and near-field coupling [64] between the patterned and continuous conductor layers. However, the size of the metamaterial absorbing cell is only 1/5–1/3 of the wavelength of the incident electromagnetic wave. In order to use the perfect metamaterial absorber more practically, the cell size to wavelength ratio should be further reduced. Generally, the existing characteristics of metamaterial absorbers have high absorptance at a specific frequency, and the narrow absorption band and polarization sensitivity also need to be improved [65].

2.1. Development of Metamaterial Absorber

Because of the controllable permittivity and permeability of metamaterials, it provides more material basis and design ideas for designing new functional devices. Landy proposed the metamaterial-based absorber, composed of three parts, the surface is a conductive electric resonator, the middle is a polyimide (FR-4) dielectric layer, and the bottom is a metal tangent. The results show that the absorber can obtain 96% absorptance at 11.48GHz. Moreover, two possible sources of the absorption mechanism of the absorber are analyzed theoretically: ohmic loss and dielectric loss. The method to obtain perfect absorption is to equip two metal parts, the top metal pattern is used to minimize the reflectivity of an incident electromagnetic wave through impedance matching. The bottom metal part is composed of a whole piece of metal plate, which mainly blocks the transmission of electromagnetic waves. Landy only preliminarily proposed and studied the single band high absorption metamaterial absorber structure at GHz band. In order to expand the absorption bandwidth of the absorber and reduce its limitations in practical application, a metamaterial absorber working in the GHz band with dual-band absorption ability is proposed by Li [66]. Subsequently, GHz multi-band and even broadband metamaterial absorber structures have been designed and prepared.

With the development of nanotechnology, the research of metamaterial absorber has rapidly developed from microwave to terahertz. Bingham experimentally proved the terahertz metamaterial absorber; when the incident electric field is parallel to the central axis of the electric open resonance ring in the direction, the absorptance of the absorber measured experimentally reaches 70%, which is in good agreement with the simulation value of 68%. When the incident electric field is perpendicular to the central axis of the electric open resonance ring, the results show that the absorptance of the absorber is only 27%. It can be seen that the absorber structure is polarization-dependent because the metamaterial structure is a double symmetrical structure. Therefore, to obtain a polarization-insensitive terahertz perfect metamaterial absorber, it is necessary to design a metamaterial structure with quadruple rotational symmetry [68-70].

However, only single band absorption is limited in practical application; researchers began to focus on the metamaterial absorbers insensitive to polarization, multi, or broadband absorption. Landy proposed a metamaterial absorber that achieves perfect absorption at 1.13 THz and has polarization-insensitive characteristics [71]. The metamaterial absorber designed by Shen is a three-layer structure with triple-band absorption. The top structure is three square metal rings nested together, the dielectric layer is PCB board with a thickness of 0.78mm, and the copper plate is the bottom material with a thickness of 18μm. The absorptance of the three absorption frequency points of the absorber are 99%, 96%, and 98%, respectively, the half-height width of each absorption peak is 0.16GHz, 0.23GHz, and 0.33GHz respectively, and the higher the absorption frequency, the wider the half peak bandwidth [72]. Ye et al. developed a broadband metamaterial absorber using multi-band response combination. The characteristic of the absorber is polarization insensitive. The structure of a metamaterial absorber is a periodic structure with a metal cross, which has a total of seven layers, including three cross metal resonance units of different sizes, three dielectric layers and one metal plate. The Keding metal cutting line inspires the design of the metal cross structure, but the metal cutting line structure does not have 90° rotation symmetry, so it does not have polarization insensitivity [73]. The concentric ring structure was proposed, and the multi-band absorption was obtained [74]. Yao proposed an elliptical metamaterial absorber, which uses graphene material and obtains double band absorption in the terahertz band [75].

Many similar resonance responses are combined, and the absorption peaks are superimposed to realize broadband absorption finally. Ma et al. designed an ultra-wideband absorber with a multilayer structure. The results show that the bandwidth of absorptance over 90% is 1.07THz [76]. Cui theoretically proposed an ultra-wideband thin-film infrared absorber made of "sawtooth" anisotropic metamaterials. The absorptance within the absorption bandwidth is higher than 95% [77]. Li et al. proposed a broadband metamaterial absorber in GHz, achieving broadband absorption of 100 GHz. The absorber structure is three I-shaped resonators arranged in parallel in the horizontal direction [78]. Munaga et al. designed a broadband metamaterial absorber in the split loop unit using resistance [79]. Shi et al. proposed a broadband metamaterial absorber designed in the microwave that can achieve wide-angle incidence, and the absorption exceeds 80% when the electromagnetic wave is vertically incident within 6~16.2GHz [80]. Ghobadi et al. provided a nearly perfect metamaterial absorber with a metal-dielectric layer, which has good polarization-insensitive broadband microwave absorption performance in the visible frequency [81]. Shen et al. designed a single-layer absorber based on a multi-ring flexible structure, which has broadband absorption performance. Four metal rings, an intermediate dielectric layer, and a metal bottom plate are used to form the structure. After optimizing the structural parameters, the coupling between rings is strengthened, the resonance peaks converge, and the broadband absorption in the THz band is realized [82].

With the further development of device technology, the metamaterial absorber has expanded from terahertz frequency to infrared and visible light. Liu successfully obtained an infrared metamaterial absorber in theory and experiment. The metamaterial structure device is on a silicon substrate, and it has experimentally demonstrated the absorption of spatial and frequency changes by using two different metamaterial sublattices, which may be applied in many aspects, including high spectral sub sampling imaging [38]. Hao et al. proposed a metamaterial absorber composed of a group of square metal structures and pointed out that the absorber is an ultra-thin, wide-angle, sub-wavelength, and high-performance in the visible frequency range. The simulation results show that the absorber can obtain almost perfect absorption. The experimental results show that 88% absorption was achieved [83].

2.2. Applications of Metamaterial Absorber

The application of the metamaterial absorber reported in the recent reference is summarized in Table I. Xiong et al. proposed the absorption tunable broadband metamaterial absorber based on Dirac semimetal. It is found that the working frequency band can change from narrowband to broadband under different top patterns. In addition, by changing the Fermi level of graphene in the structure, the bandwidth with perfect absorption (the absorptance over 90%) can be dynamically adjusted up to 3.4 THz at the highest [84]. Zhang and his colleagues design a terahertz multiple broadband absorber that can be absorption tunable. The results indicate there are six band absorptances over 90%, ranging from 0 to 9.5 THz. Depending on the characteristic that the conductivity change of VO_2 will lead to the material's phase change, the structure achieves absorption tunable with an amplitude as high as 93% [85]. Wang proposed a terahertz metamaterial absorber made of graphene as the main material. Compared with the absorber proposed in the previous article, it upgraded one absorption tunable broadband to two, and the absorptance of these two broadbands can be adjusted independently, further expanding the application range [86]. Appasani proposed a terahertz metamaterial absorber that can be used as a temperature sensor and refractive index detector. The core component of temperature sensing is InSb substrate. Depending on the characteristics that the relative dielectric constant of this material will change at different temperatures, this structure can be used as a temperature sensor.

At the same time, when the ambient temperature is 300K, different refractive indexes have a good linear relationship with the resonance point, so the structure can also be used as a refractive index detector [87]. Gao et al. put forward a unique broadband metamaterial absorber. The air and water layers were introduced into the structure, which absorbed electromagnetic waves and weakened infrared radiation. The absorber has a broad bandwidth in the microwave band, and the perfect absorption starts from 14.4 GHz to 25 GHz [88]. Lan et al. also designed a completely different manufacturing method for the metamaterial absorber, which is different from the superposition of different substances with the traditional sandwich structure. This method relies on 3D printing and encapsulates the designed water array structure in the shell, making manufacturing easier. The perfect absorption starts at the frequency of 0.3 THz, and the bandwidth reaches 1.2 THz. At the same time, according to the analysis and calculation of the permittivity of water in the normal temperature range, it is demonstrated that the absorber has thermal stability [89]. Liu et al. demonstrated a completely symmetrical metamaterial absorber. The symmetrical structure makes the absorber polarization insensitive. The absorption bandwidth of the absorber up to 3.26THz, starting from 4.22THz. In addition to changing the geometric parameters of the structure, adjusting the chemical potential energy of graphene can also shift the absorption band [90]. Luo broke the barrier between GHz and THz and proposed a metamaterial absorber that can simultaneously realize the GHz-band and THz-band absorption [91].

Luo et al. proposed a broadband metamaterial absorber with switchable bandwidth, taking advantage of the characteristic that VO2 can switch between the metal and insulating phase through external temperature change. When VO_2 is in the insulating phase, the absorber realizes perfect absorption at 1.75 THz, and the bandwidth is 0.71 THz. When the temperature of the external environment is raised, VO_2 becomes a metal phase. At this time, the absorber can begin to achieve ultra-broadband perfect absorption at 2.6 THz, and the bandwidth is up to 4.4 THz [92]. Nie proposed a broadband absorber, which achieves perfect absorption in the bandwidth range from 1.03 THz to 1.63 THz. Different from absorbers used as terahertz switches, temperature sensors, and other applications, the absorber can be used to detect the concentration of pesticide Chlorpyrifos. When the Chlorpyrifos is placed on the surface of the absorber, it is found that with the increase of the concentration of Chlorpyrifos, the resonance point of the absorber will shift to a high frequency, and the absorptance of the resonance point will decrease, so it can be used as the detection of Chlorpyrifos concentration [93].

Table 1. Application of recently reported metamaterial absorber

Ref	Working Band	Main Function or Characteristic
84	THz	Changing the Fermi level of graphene can make the absorber absorption tunable
85	THz	The absorptance varies with VO_2 conductivity
86	THz	Double broadband absorption, and absorptance can be adjusted separately
87	THz	Can be used to detect temperature and refractive index
88	GHz	Composed of air layer and water layer, it has optical transparency and can significantly reduce infrared radiation
89	THz	Has thermal stability and is insensitive to polarization angle and incident angle
90	THz	Changing the chemical potential can make the absorber absorption tunable
91	GHz and THz	Realize the absorption in GHz-band and THz-band simultaneously
92	THz	Dual broadband absorption, which changes the absorption bandwidth by changing the temperature
93	THz	Detect the concentration of pesticide Chlorpyrifos

3. Theory of Absorption

A metamaterial absorber is a metamaterial based on electromagnetic resonance. For achieving perfect absorption, the core is to minimize the reflectivity of incident electromagnetic wave and block the transmission of an electromagnetic wave through the bottom continuous metal plate to realize the perfect absorption through matching material input impedance and free space impedance. Furthermore, the frequency can be adjusted by artificially designing different structures and sizes. This section introduces the theoretical basis of electromagnetic metamaterial absorber, mainly including impedance matching theory [102], multiple reflection interference theory [63], and equivalent circuit theory [94]. These theoretical foundations and numerical simulation methods lay a foundation for the subsequent research on designing and applying electromagnetic tunable metamaterial absorbers.

3.1. Impedance Matching Theory

Impedance matching theory is the basic theory for analyzing the performance of metamaterial absorbers. By artificially controlling the dielectric constant and permeability of metamaterial structure, the input impedance of a material is matched with the impedance of free space to minimize the reflectivity of the absorber so that the incident electromagnetic wave can enter into the absorber and be dissipated as much as possible, and the absorber structure can obtain the maximum absorptance. When the scattering of incident electromagnetic wave and surface wave can be ignored, the absorptance $A(\omega)$, reflectivity $R(\omega)$, and transmissivity $T(\omega)$ of the absorber satisfies equation (1). Therefore, to achieve high absorptance, the reflectivity $R(\omega)$ and transmissivity $T(\omega)$ must be minimized. The size of the reflection coefficient S_{11} is determined by the matching degree of the input impedance of the material and the impedance of the free space, which can be expressed by the following:

$$S_{11} = \frac{Z_r - Z_0}{Z_r + Z_0} \tag{2}$$

Z_r and Z_0 are the input impedance of material and the impedance of free space, respectively, which can be expressed as:

$$Z_r = \sqrt{\frac{\mu_r}{\varepsilon_r}} \tag{3}$$

$$Z_0 = \sqrt{\frac{\mu_0}{\varepsilon_0}} \tag{4}$$

(ε_r, μ_r) is the permittivity and permeability of free space, (ε_0, μ_0) is the dielectric constant and permeability of the material.

The transmission coefficient S_{21} reflects the energy loss of the incident electromagnetic wave when passing through the metamaterial absorber, which can be expressed by the following:

$$S_{21}^{-1} = \left[\sin(nkd) - \frac{i}{2}(Z + \frac{1}{Z})\cos(nkd)\right]e^{ikd} \tag{5}$$

where Z is the equivalent complex impedance, n is the refractive index, k is the wave vector and d is the thickness of the medium, which can be expressed by the following

$$Z = \frac{Z_r}{Z_0} = Z_1 + iZ_2 \tag{6}$$

$$n = n_1 + in_2 \tag{7}$$

$$k = \omega / c \tag{8}$$

c is velocity of light in vacuum, ω is the angular frequency.

Two conditions need to be met to achieve perfect absorption: the material's impedance perfectly should match the impedance of the free space wave, and the imaginary part of the refractive index tends to infinity. At this time, $Z_r = Z_0$ and $Z = 1$. Combined with the above formula, the reflectivity and transmissivity meet the following

$$R(\omega) = |S_{11}|^2 = \left[\frac{Z_r - Z_0}{Z_r + Z_0}\right]^2 = 0 \tag{9}$$

$$T(\omega) = |S_{21}|^2 = e^{-2kdn_2} \tag{10}$$

$A(\omega) \approx 1$, means the perfect absorption.

3.2. Equivalent Circuit Theory

The impedance matching theory can only qualitatively but not quantitatively analyze the absorption performance of the absorber. The equivalent circuit theory, which holds the structural elements, can be expressed as a resonance circuit composed of equivalent capacitance and equivalent inductance, and their values can be solved by using the geometric parameters of the metal pattern layer. Then the resonance frequency of the metamaterial absorber can be approximately calculated by using the equivalent inductance and equivalent capacitance. Also, changing the geometric parameters of the metal pattern layer can predict the systematic changes.

Zhou et al. calculated the resonance frequency of metamaterial absorbers by equivalent circuit theory [95]. Two tangents of the absorber unit cell are placed on both sides of the dielectric layer to form a tangent pair, and four tangents are placed in two pairs on both sides of the dielectric layer to form two tangent pairs. It is known that the inductance of the parallel plate is $L = \mu_0(l*t_s)/W$, where μ_0 is the permeability in a vacuum, l is the length of the tangent, W is the width, t_s is the spacing between short conductors, and the capacitance of the tangent pair is $C = \varepsilon_r\varepsilon_0(l*W)/4t_s$, where ε_0 is the dielectric constant in vacuum, ε_r is the relative dielectric constant of the tangent region, and the magnetic resonance frequency f_m is:

$$f_m = \frac{1}{2\pi\sqrt{LC}} \approx \frac{1}{\pi l\sqrt{\varepsilon_r\varepsilon_0\mu_0}} = \frac{c}{\pi l\sqrt{\varepsilon_r}} \tag{11}$$

c is the velocity of light in a vacuum. The above formula shows that the resonance frequency f_m is inversely proportional to the length L of the tangent pair but independent of the width W.

Lam et al. also used the theory to explain the relationship between magnetic resonance frequency and tangent width W in tangent pair medium [96]. In order to obtain the value of equivalent capacitance (C_{eff}) and equivalent inductance (L_{eff}), the differential equations of charge density and current density are established, respectively, and the total induced charge and current are obtained by integration. Finally, the equivalent capacitance and inductance can be obtained by:

$$U_E = \frac{1}{2}\int \varepsilon E^2 dv = \frac{Q^2}{2C_{eff}} \tag{12}$$

$$U_B = \frac{1}{2}\int \frac{B^2}{\mu} dv = \frac{1}{2}L_{eff}I^2 \tag{13}$$

The U_E and U_B of the above electric resonance and magnetic resonance can be obtained by assuming that the current in the tangent pair flows in the same direction and the opposite direction, respectively. Then the frequency of electric resonance and magnetic resonance can be obtained. Lam et al. calculated the relationship between the frequency of electric resonance and magnetic resonance and the distance between tangent pairs [97]. Pang et al. also applied the equivalent circuit theory to the metamaterial absorber to clarify the influence of the absorber's surface resistance on the absorber's frequency [98]. In the calculation process, the equivalent impedance of the metal pattern layer on the surface of the absorber can be expressed by *LC* circuit. The results show that narrowband or broadband absorbing absorbers can be obtained by changing the surface impedance, and this method can guide the design of high-speed and effective metamaterial absorbers.

3.3. Multiple Reflection Interference Theory

The impedance matching theory is used to analyze the absorptance of the absorber qualitatively but not quantitatively. Although the equivalent circuit theory quantitatively analyzes the absorption performance of the absorber, it is only suitable for the simple absorber model, and it is not easy to give the equivalent circuit model for a complex absorber structure. Chen proposed the multiple reflection interference theory, which can not only quantitatively analyze the absorption mechanism of the absorber but also explain the interaction between the metal layers in the absorber and is suitable for complex structures and oblique incidence of electromagnetic waves [63]. Take the common sandwich structure as an example: the surface metal resonance layer, the intermediate dielectric layer, and the bottom metal plate. The multiple reflection interference model is shown in Figure 2. The theory points out that in the interference model, the two metal layers in the metamaterial absorber can be regarded as an impedance tunable surface with zero thickness, and for the simple interference model, it is assumed that the near-field coupling between the surface metal pattern layer and the ground plane can be ignored [67]. It can be seen from Figure 2 that when the electromagnetic wave is obliquely incident on the surface of metamaterial at an angle of α, the incident electromagnetic wave at the interface between the air and the dielectric layer can be divided into two parts: one is reflected back into the air, and the reflection coefficient is $S_{11}e^{i\theta}_{11}$, another is transmitted into the absorber, and the transmission coefficient is $S_{21}e^{i\theta}_{21}$. The incident electromagnetic wave continues to propagate forward to interface 2 with the propagation phase β, and interface 2 is the metal plate at the bottom layer, which is an ideal conductor. The electromagnetic wave propagating above is fully reflected with the coefficient S_{23} = -1. The incident electromagnetic wave continues to propagate forward to interface 2 between the air and the dielectric layer. At this time, the propagation phase is still β, and the electromagnetic wave also produces reflection with the coefficient of $S_{22}e^{i\varphi}_{22}$ and transmission with the coefficient of $S_{21}e^{i\theta}_{21}$. Therefore, the reflected wave of the whole absorber can be obtained by the superposition of all the above-reflected waves:

$$
\begin{aligned}
S_{11total} &= S_{11} + S_{21}e^{i\beta}S_{23}e^{i\beta}S_{12} + S_{21}e^{i\beta}S_{23}e^{i\beta}(S_{22}e^{i\beta}S_{23}e^{i\beta})S_{12} \\
&+S_{21}e^{i\beta}S_{23}e^{i\beta}(S_{22}e^{i\beta}S_{23}e^{i\beta})^2 S_{12} + S_{21}e^{i\beta}S_{23}e^{i\beta}(S_{22}e^{i\beta}S_{23}e^{i\beta})^3 S_{12} + \ldots + \\
&S_{21}e^{i\beta}S_{23}e^{i\beta}(S_{22}e^{i\beta}S_{23}e^{i\beta})^n S_{12} \\
&= S_{11} + S_{12}S_{21}S_{23}e^{i2\beta}\sum_{n=0}^{\infty}(S_{22}S_{23}e^{i2\beta})^n \\
&= S_{11} + \frac{S_{12}S_{21}S_{23}e^{i2\beta}}{1 - S_{22}S_{23}e^{i2\beta}}
\end{aligned}
\tag{14}
$$

where the propagation phase in the dielectric layer is $\beta = kd$, k is the wave number propagating in the dielectric layer, and d is the distance propagating from the dielectric layer. The expressions are

$$k = \frac{2\pi}{\lambda} = 2\pi\sqrt{\varepsilon_2\mu_2}\frac{f}{c_0} \tag{15}$$

$$d = \frac{t}{\cos\beta} = \frac{t}{\cos(\arcsin(\sin\alpha / \sqrt{\varepsilon_2\mu_2 / \varepsilon_1\mu_1}))} \tag{16}$$

$$\beta = kd = \frac{2\pi\sqrt{\varepsilon_2\mu_2}ft}{\cos(\arcsin(\sin\alpha / \sqrt{\varepsilon_2\mu_2 / \varepsilon_1\mu_1}))c_0} \tag{17}$$

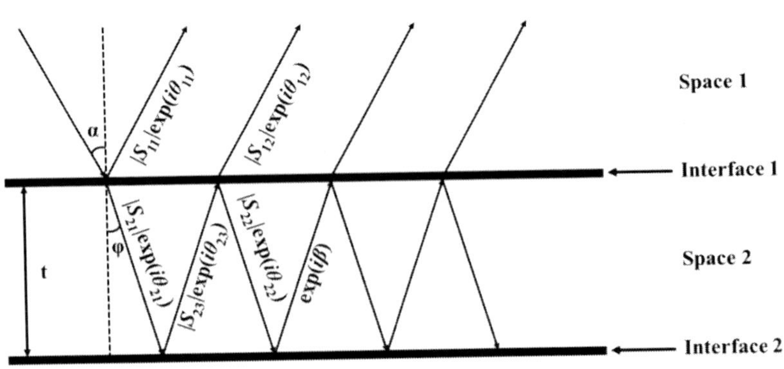

Figure 2. Interference model and variables of metamaterial absorber.

In order to simplify the calculation, it always assumes that the dielectric layer is FR4, the electromagnetic wave is incident vertically, and the bottom is metal plate. When space 1 is air and space 2 is FR4, there is $\varepsilon_1=\mu_1=\mu_2=1$. When the electromagnetic wave is incident vertically, $\alpha = 0$. The bottom layer is a continuous metal plate means $S_{23} = 1*e^{i\pi} = -1$. Substituting the three assumptions into (1-14) - (1-16) can obtain:

$$\beta = kd = \frac{2\pi\sqrt{\varepsilon_2}ft}{c_0} \quad (18)$$

$$S_{11total} = S_{11} - \frac{S_{12}S_{21}e^{i2\beta}}{1+S_{22}e^{i2\beta}} \quad (19)$$

The absorptance of the metamaterial absorber is $A(\omega) = 1\text{-}R(\omega) = 1\text{-}|S_{11total}|^2$, so it is necessary for the absorber to obtain perfect absorption, that is, $A(\omega) = 1$, $S_{11total} = 0$, the following phase conditions and amplitude conditions must be met by

$$|S_{11}| = |S_{12}S_{21} - S_{11}S_{22}| \quad (20)$$

$$\phi(S_{11}) - \phi(S_{12}S_{21} - S_{11}S_{22}) - 2\beta = 2m\pi, |m| = 0,1,2\ldots \quad (21)$$

4. Research Prospect on Metamaterial Absorber

It has been more than a decade since the metamaterial absorber was first proposed and widely studied, but some problems still exist [99]. In terms of absorption performance, because the loss of materials decreases with the decrease of frequency, it is challenging to realize broadband absorption in a low-frequency band. In terms of function, with the increasing complexity of application scenarios, multifunctional integrated metamaterials will gradually become the focus of attention. In terms of structure, the two-dimensional metamaterial is very sensitive to oblique incident electromagnetic waves. The new three-dimensional structure is expected to become a new breakthrough in the research of metamaterial absorbers.

4.1. Low Frequency Ultra-Thin Broadband Metamaterial Absorbers

The properties and functions of electromagnetic waves in different frequencies are different. Due to the long wavelength in the low-frequency band, the volume of the corresponding traditional absorbing structure is large, and the relationship between thickness and bandwidth restricts the design of ultra-thin broadband metamaterial absorbers [100]. Especially in the low-frequency band, limited by Snoek theory, while the absorptance of metamaterial in the low-frequency band is poor, which is difficult to meet the current demand for low-frequency ultra-thin metamaterials with single metamaterial structure [101]. Low-frequency broadband absorption can be realized through combining traditional magnetic absorbing materials and metamaterials. For example, Mou designed a broadband absorber loaded with Non-Foster circuits, which can work in the frequency range of 150 to 950 MHz [103]. Later, in order to reduce the thickness, magnetic materials were used by Mou et al. instead of dielectric materials, finally reduced the thickness from 30mm to 7.5mm [104]. In 2018, Banadaki et al. reduced the minimum operating frequency by 20% on the base of the reference structure by modifying the unit structure of the absorber. They achieved an absorptance of over 90% in the range of 1.35 GHz to 3.5 GHz, which greatly improved the practical application ability of the absorber [105]. Common methods to realize low-frequency wave absorption also include loading resistance film, loading passive components or active components. Li et al. designed a tunable absorber with a thickness of only 1mm based on the composite method of ferrite and metamaterials and realized absorption-tunable in the range of 0.2 GHz to 7.6 GHz. However, the bandwidth problem is still a common problem faced by metamaterials. With the development of low-frequency early warning radar, the research on low-frequency broadband absorption will be paid more attention [106]. The composite absorber combined with a square metal patch structure and resistance frequency selective surface proposed by Cheng has a thickness of only 2 mm, and the absorptance above 90% in the range of 3 GHz to 25 GHz. This structure shows excellent absorption capacity in a wide frequency range [107]. The absorber's core is a cross-structure pattern formed by carbonyl iron powder, which is different from the common method of broadening the bandwidth by adding layers of metal patterns. The structure can expand the bandwidth by changing the size of cross structures in the unit cell or combining cross structures of different sizes so that the absorption peak size and bandwidth can be tuned in the range of 1-8 GHz. Thus, a relatively wide bandwidth is obtained [108].

4.2. Multifunctional Integrated Metamaterial Absorber

Metamaterials use the accumulation of spatial phases to control electromagnetic waves and realize different functions. Due to the limitation of the early technology, researchers innovatively proposed a two-dimensional metamaterial. Placing some sub-wavelength structural pieces on the medium's surface makes it possible to achieve flexible control of the electromagnetic wave-front and polarization. In recent years, it has become a hot spot and frontier in the field of the metamaterial. Using the abrupt phase of metamaterials to realize specific functions can be summarized as follows: transmission phase regulation, geometric phase regulation, and hybrid modulation of transmission and geometric phase. Integrating multiple functions into metamaterials is the future development direction. In 2017, Zhang proposed a metamaterial with optical transparency and microwave broadband absorption [109]. Good light transmission characteristics were obtained using indium-tin-oxide (ITO) thin film, and broadband absorption was achieved in the range of 8.3 to 17.4 GHz. With the rapid development of information technology, researchers have been widely concerned about multifunctional integrated electromagnetic devices that integrate absorbers and other electromagnetic functions. In the field of metamaterials, the reflection phase can be controlled arbitrarily by changing the geometric size of the metal. In the absorber design, perfect absorption can be realized through impedance matching. However, up to now, these functions have only been realized on a single separated device. In 2019, Sun combined metal structure and loaded resistance element to realize the multifunctional metamaterial with the functions of wave-front plasticity and wave absorption in microwave band through the anisotropy of the unit [110].

Similarly, this type of absorber has also been designed in the THz band. Peng et al. designed a metamaterial unit that can realize polarization conversion and wave absorption at the same time by combining metal structure with graphene [111]. Then Chen et al. proposed a multifunctional metamaterial that can realize the conversion between super-lens and wave absorbers in the THz frequency band [112]. In addition, multi-field compatibility metamaterials, wave absorption, and transmission integration, and multi-spectrum compatibility are also expected to become a new direction for researchers to explore.

4.3. Three-Dimensional Structure Metamaterial Absorber

At the early stage of the development of metamaterials, people have carried out a lot of exploration and research on three-dimensional structure devices [113, 114]. Although the electromagnetic wave can be manipulated arbitrarily according to people's wishes through three-dimensional metamaterials, its high loss and manufacturing complexity significantly limit its application [115, 116]. Considering the advantages of three-dimensional structure in regulating electromagnetic waves and the development of 3D printing technology, it is expected to provide a new direction for metamaterials by combining them. In 2018, Xiong et al. successfully designed a broadband metamaterial absorber composed of a three-layer square matrix structure with a total thickness of 4.7 mm, which fully proved the possibility of 3D printing technology in metamaterials [117]. Compared with the two-dimensional metamaterial structure, the three-dimensional metamaterial structure can regulate the electromagnetic wave more effectively, especially in low-frequency band and miniaturization. For example, Lim proposed to use a 3D printing structure to realize stepped cross-structure metamaterial absorber in 2018 [118]. It is found that three-dimensional structure is beneficial to the miniaturization of metamaterial absorbers. In the same year, Jiang et al. presented a honeycomb structure with a resistance film as a three-dimensional metamaterial structure. The absorber can realize broadband absorption in the range of 3.53 to 24 GHz, and 3D printing technology is conducive to the lightweight metamaterial absorber structure [119]. In 2019, Xie et al. injected water into a plastic container manufactured by 3D printing to manufacture an all-dielectric metamaterial absorber in the microwave band [120]. Other 3D studies include graphene metamaterial absorbing structure and FSS absorbing structure [121, 122]. At the same time, the combination of non-planar structures and other structures also provides a new idea for realizing broadband absorption. Li et al. studied the "standing" metamaterial absorber loaded with magnetic resonance structure [123]. Because the three-dimensional structure introduces the regulation of spatial angle, it is expected to be applied to multifunctional integrated metamaterials, which is conducive to promoting the research and application of new multifunctional equipment. In 2019, Chen designed a tunable and multifunctional three-dimensional metamaterial structure with origami technology [124]. Switching between functions such as absorber, specular reflector, and inverse reflector can be realized by adjusting the apex angle. But there is a need for further research on how to accurately and automatically adjust the apex angle to make tuning more reliable in the experiment and practical application. The introduction of three-dimensional structure brings new vitality to developing new metamaterial devices. With the progress of technology, more new structures will appear in the future, and people will become freer to regulate and control electromagnetic waves.

Conclusion

In this chapter, metamaterial absorbers are introduced in detail. The first section starts with metamaterials, compares the characteristics, differences, and advantages between metamaterials and traditional materials, and emphasizes the special properties that metamaterials do not have compared with traditional materials. The metamaterial absorber is introduced through many examples of the application of metamaterials. First, the structural physics principle of metamaterial absorbers is introduced in detail, the development process of metamaterial absorbers from its first proposal by Landy in 2008 to recent years is summarized, and examples are given to illustrate the different application scenarios of metamaterial absorbers. Then, three common methods to optimize and improve the absorption properties of metamaterial absorbers are discussed in detail: impedance matching theory, equivalent circuit theory, and multiple reflection interference theory. Among them, the impedance matching theory is used to analyze the absorptance qualitatively, and the equivalent circuit theory is used to analyze the absorptance quantitatively. However, these two methods are mainly suitable for simple models, and multiple reflection interference theory is suitable for complex models. Finally, in the second part of the chapter, the main content is the prospect of metamaterial absorbents. This part discusses three problems encountered in the current research: the realization of low-frequency ultra-broadband absorption, multi-function integrated metamaterial, and new three-dimensional metamaterial. Besides, some current research results around these problems are briefly described. It provides a reference direction for readers in the future research of metamaterial absorbers.

References

[1] De Rosa, Igor Maria, Fabrizio Sarasini, Maria Sabrina Sarto, and Alessio Tamburrano. "EMC impact of advanced carbon fiber/carbon nanotube reinforced

composites for next-generation aerospace applications." *IEEE Transactions on Electromagnetic Compatibility* 50, no. 3 (2008): 556-563.

[2] Baev, Alexander, Marek Samoc, Paras N. Prasad, Mykhaylo Krykunov, and Jochen Autschbach. "A quantum chemical approach to the design of chiral negative index materials." *Optics Express* 15, no. 9 (2007): 5730-5741.

[3] Kaur, Sandeep, Neil J. Findlay, Alexander L. Kanibolotsky, Saadeldin ET Elmasly, Peter J. Skabara, Rory Berridge, Claire Wilson, and Simon J. Coles. "Electrochromic properties of a poly (dithienylfuran) derivative featuring a redox-active dithiin unit." *Polymer Chemistry* 3, no. 8 (2012): 2277-2286.

[4] Sun, Mengxiao, Xuliang Lv, Aming Xie, Wanchun Jiang, and Fan Wu. "Growing 3D ZnO nano-crystals on 1D SiC nanowires: enhancement of dielectric properties and excellent electromagnetic absorption performance." *Journal of Materials Chemistry C* 4, no. 38 (2016): 8897-8902.

[5] Markoš, Peter, and Costas M. Soukoulis. "Transmission properties and effective electromagnetic parameters of double negative metamaterials." *Optics express* 11, no. 7 (2003): 649-661.

[6] Lapine, M., and S. Tretyakov. "Contemporary notes on metamaterials." *IET microwaves, antennas & propagation* 1, no. 1 (2007): 3-11.

[7] Smith, David R., John B. Pendry, and Mike CK Wiltshire. "Metamaterials and negative refractive index." *Science* 305, no. 5685 (2004): 788-792.

[8] Seddon, N., and T. Bearpark. "Observation of the inverse Doppler effect." *Science* 302, no. 5650 (2003): 1537-1540.

[9] Pendry, John Brian. "Negative refraction makes a perfect lens." *Physical review letters* 85, no. 18 (2000): 3966.

[10] Walser, Rodger M. "Electromagnetic metamaterials." In *Complex Mediums II: beyond linear isotropic dielectrics*, vol. 4467, pp. 1-15. SPIE, 2001.

[11] Shelby R A, Smith D R, Schultz S, Experimental verification of a negative index of refraction. Science 2001, 292(5514): 77-79.

[12] Ebbesen, Thomas W., Henri J. Lezec, H. F. Ghaemi, Tineke Thio, and Peter A. Wolff. "Extraordinary optical transmission through sub-wavelength hole arrays." *nature* 391, no. 6668 (1998): 667-669.

[13] Yannopapas, Vassilios, Emmanuel Paspalakis, and Nikolay V. Vitanov. "Electromagnetically induced transparency and slow light in an array of metallic nanoparticles." *Physical Review B* 80, no. 3 (2009): 035104.

[14] Lu, Yuehui, X. R. Jin, H. Y. Zheng, Y. P. Lee, J. Y. Rhee, and W. H. Jang. "Electromagnetically-induced Transparency in Metamaterials Based on the Second-order Magnetic Plasmon Resonance." *Journal of the Korean Physical Society* 58, no. 4 (2011): 973-976.

[15] Schurig, David, Jack J. Mock, B. J. Justice, Steven A. Cummer, John B. Pendry, Anthony F. Starr, and David R. Smith. "Metamaterial electromagnetic cloak at microwave frequencies." *Science* 314, no. 5801 (2006): 977-980.

[16] Liu, Xiaoming, Qian Zhao, Chuwen Lan, and Ji Zhou. "Isotropic Mie resonance-based metamaterial perfect absorber." *Applied Physics Letters* 103, no. 3 (2013): 031910.

[17] Liu YH, Luo CR, and Zhao XP. "The left-handed material of H-type structural unit with negative dielectric constant and permeability is realized at the same time." *Acta Phys. Sin.* 2007, 56(10): 5883-5889.

[18] Zhang D K, Zhang Y W, Hao L, et al. "Experimental study on the characteristics of one-dimensional metamaterials constructed by lumped L-C elements." *Acta Phys. Sin.* 2005, 54(2): 768-772.

[19] Zhou, Lei, Hongqiang Li, Yaqin Qin, Zeyong Wei, and C. T. Chan. "Directive emissions from subwavelength metamaterial-based cavities." *Applied Physics Letters* 86, no. 10 (2005): 101101.

[20] Tao, Jiang, Cui Wan-Zhao, Ma Wei, Yuan Yu, Wang Dong-Xing, and Ran Li-Xin. "High directive cavity antenna based on one-dimensional LHM-RHM resonator." *Chinese Physics Letters* 26, no. 10 (2009): 104101.

[21] Shen, Xiaopeng, and Tie Jun Cui. "Photoexcited broadband redshift switch and strength modulation of terahertz metamaterial absorber." *Journal of Optics* 14, no. 11 (2012): 114012.

[22] Shalaev, V. M., T. A. Klar, V. P. Drachev, and A. V. Kildishev. "Optical Negative-Index Metamaterials: From Low to No Loss." *Photonic Metamaterials: From Random to Periodic*.

[23] Driscoll, Tom, Hyun-Tak Kim, Byung-Gyu Chae, Bong-Jun Kim, Yong-Wook Lee, N. Marie Jokerst, Sabarni Palit, David R. Smith, Massimiliano Di Ventra, and Dimitri N. Basov. "Memory metamaterials." *Science* 325, no. 5947 (2009): 1518-1521.

[24] Liu, Na, Hongcang Guo, Liwei Fu, Stefan Kaiser, Heinz Schweizer, and Harald Giessen. "Three-dimensional photonic metamaterials at optical frequencies." *Nature materials* 7, no. 1 (2008): 31-37.

[25] Tuong, P. V., Y. J. Yoo, J. W. Park, Y. J. Kim, K. W. Kim, Y. H. Kim, H. Cheong, L. Y. Chen, and Y. P. Lee. "Multi-plasmon-induced perfect absorption at the third resonance in metamaterials." *Journal of Optics* 17, no. 12 (2015): 125101.

[26] Wang, Xiaonong, Chunrong Luo, Gang Hong, and Xiaopeng Zhao. "Metamaterial optical refractive index sensor detected by the naked eye." *Applied Physics Letters* 102, no. 9 (2013): 091902.

[27] Appasani B. "Temperature Tunable Seven Band Terahertz Metamaterial Absorber Using Slotted Flower–Shaped Resonator on an InSb Substrate." *Plasmonics* (2021): 16: 833-839.

[28] Hendrickson, Joshua, Junpeng Guo, Boyang Zhang, Walter Buchwald, and Richard Soref. "Wideband perfect light absorber at midwave infrared using multiplexed metal structures." *Optics letters* 37, no. 3 (2012): 371-373.

[29] Liu, Na, Martin Mesch, Thomas Weiss, Mario Hentschel, and Harald Giessen. "Infrared perfect absorber and its application as plasmonic sensor." *Nano letters* 10, no. 7 (2010): 2342-2348.

[30] Niesler, Fabian BP, Justyna K. Gansel, Sarah Fischbach, and Martin Wegener. "Metamaterial metal-based bolometers." *Applied Physics Letters* 100, no. 20 (2012): 203508.

[31] Wang, Bingnan, Koon Hoo Teo, Tamotsu Nishino, William Yerazunis, John Barnwell, and Jinyun Zhang. "Experiments on wireless power transfer with metamaterials." *Applied Physics Letters* 98, no. 25 (2011): 254101.

[32] Aydin, Koray, Vivian E. Ferry, Ryan M. Briggs, and Harry A. Atwater. "Broadband polarization-independent resonant light absorption using ultrathin plasmonic super absorbers." *Nature communications* 2, no. 1 (2011): 1-7.

[33] Hedayati, Mehdi Keshavarz, Franz Faupel, and Mady Elbahri. "Tunable broadband plasmonic perfect absorber at visible frequency." *Applied Physics A* 109, no. 4 (2012): 769-773.

[34] Hall, Anthony Shoji, Muhammad Faryad, Greg D. Barber, Liu Liu, Sema Erten, Theresa S. Mayer, Akhlesh Lakhtakia, and Thomas E. Mallouk. "Broadband light absorption with multiple surface plasmon polariton waves excited at the interface of a metallic grating and photonic crystal." *ACS nano* 7, no. 6 (2013): 4995-5007.

[35] Munk, Ben A. *Frequency selective surfaces: theory and design*. John Wiley & Sons, 2005.

[36] Engheta, Nader. "Thin absorbing screens using metamaterial surfaces." In *IEEE Antennas and Propagation Society International Symposium (IEEE Cat. No. 02CH37313)*, vol. 2, pp. 392-395. IEEE, 2002.

[37] Landy, N. I冊, S. Sajuyigbe, Jack J. Mock, David R. Smith, and Willie J. Padilla. "Perfect metamaterial absorber." *Physical review letters* 100, no. 20 (2008): 207402.

[38] Liu, Xianliang, Tatiana Starr, Anthony F. Starr, and Willie J. Padilla. "Infrared spatial and frequency selective metamaterial with near-unity absorbance." *Physical review letters* 104, no. 20 (2010): 207403.

[39] Liu, Xianliang, Talmage Tyler, Tatiana Starr, Anthony F. Starr, Nan Marie Jokerst, and Willie J. Padilla. "Taming the blackbody with infrared metamaterials as selective thermal emitters." *Physical review letters* 107, no. 4 (2011): 045901.

[40] Hao, Jiaming, Lei Zhou, and Min Qiu. "Nearly total absorption of light and heat generation by plasmonic metamaterials." *Physical review B* 83, no. 16 (2011): 165107.

[41] Shannon, Steven, James Paul Holloway, Kirk Flippo, and Mary L. Brake. "A spatially resolved optical emission sensor for plasma etch monitoring." *Applied physics letters* 71, no. 11 (1997): 1467-1468.

[42] Ramahi, Omar M., Thamer S. Almoneef, Mohammed AlShareef, and Muhammed S. Boybay. "Metamaterial particles for electromagnetic energy harvesting." *Applied Physics Letters* 101, no. 17 (2012): 173903.

[43] Qi, Mei Qing, Wen Xuan Tang, Hui Feng Ma, Bai Cao Pan, Zui Tao, Yong Zhi Sun, and Tie Jun Cui. "Suppressing side-lobe radiations of horn antenna by loading metamaterial lens." *Scientific reports* 5, no. 1 (2015): 1-6.

[44] Chiam, Sher-Yi, Ranjan Singh, Weili Zhang, and Andrew A. Bettiol. "Controlling metamaterial resonances via dielectric and aspect ratio effects." *Applied Physics Letters* 97, no. 19 (2010): 191906.

[45] Li, Le-Wei, Ya-Nan Li, Tat Soon Yeo, Juan R. Mosig, and Olivier JF Martin. "A broadband and high-gain metamaterial microstrip antenna." *Applied Physics Letters* 96, no. 16 (2010): 164101.

[46] Yoo, Y. J., H. Y. Zheng, Y. J. Kim, J. Y. Rhee, J-H. Kang, K. W. Kim, H. Cheong, Y. H. Kim, and Y. P. Lee. "Flexible and elastic metamaterial absorber for low frequency, based on small-size unit cell." *Applied physics letters* 105, no. 4 (2014): 041902.

[47] Burgos, Stanley P., Rene De Waele, Albert Polman, and Harry A. Atwater. "A single-layer wide-angle negative-index metamaterial at visible frequencies." *Nature Materials* 9, no. 5 (2010): 407-412.

[48] Yoo, Young Joon, Young Joo Kim, Pham Van Tuong, Joo Yull Rhee, Ki Won Kim, Won Ho Jang, Y. H. Kim, H. Cheong, and YoungPak Lee. "Polarization-independent dual-band perfect absorber utilizing multiple magnetic resonances." *Optics express* 21, no. 26 (2013): 32484-32490.

[49] Park, Jin Woo, Pham Van Tuong, Joo Yull Rhee, Ki Won Kim, Won Ho Jang, Eun Ha Choi, Liang Yao Chen, and YoungPak Lee. "Multi-band metamaterial absorber based on the arrangement of donut-type resonators." *Optics express* 21, no. 8 (2013): 9691-9702.

[50] Pitchappa, Prakash, Chong Pei Ho, You Qian, Lokesh Dhakar, Navab Singh, and Chengkuo Lee. "Microelectromechanically tunable multiband metamaterial with preserved isotropy." *Scientific Reports* 5, no. 1 (2015): 1-10.

[51] Zang, XiaoFei, Cheng Shi, Lin Chen, Bin Cai, YiMing Zhu, and SongLin Zhuang. "Ultra-broadband terahertz absorption by exciting the orthogonal diffraction in dumbbell-shaped gratings." *Scientific reports* 5, no. 1 (2015): 1-6.

[52] Ji, Dengxin, Haomin Song, Xie Zeng, Haifeng Hu, Kai Liu, Nan Zhang, and Qiaoqiang Gan. "Broadband absorption engineering of hyperbolic metafilm patterns." *Scientific reports* 4, no. 1 (2014): 1-7.

[53] Kim, Young Ju, Young Joon Yoo, Ki Won Kim, Joo Yull Rhee, Yong Hwan Kim, and YoungPak Lee. "Dual broadband metamaterial absorber." *Optics express* 23, no. 4 (2015): 3861-3868.

[54] Cao, Tun, Chen-wei Wei, Robert E. Simpson, Lei Zhang, and Martin J. Cryan. "Broadband polarization-independent perfect absorber using a phase-change metamaterial at visible frequencies." *Scientific reports* 4, no. 1 (2014): 1-8.

[55] Li, Sucheng, Jie Luo, Shahzad Anwar, Shuo Li, Weixin Lu, Zhi Hong Hang, Yun Lai, Bo Hou, Mingrong Shen, and Chinhua Wang. "An equivalent realization of coherent perfect absorption under single beam illumination." *Scientific reports* 4, no. 1 (2014): 1-6.

[56] Guo, Yinghui, Yanqin Wang, Mingbo Pu, Zeyu Zhao, Xiaoyu Wu, Xiaoliang Ma, Changtao Wang, Lianshan Yan, and Xiangang Luo. "Dispersion management of anisotropic metamirror for super-octave bandwidth polarization conversion." *Scientific Reports* 5, no. 1 (2015): 1-7.

[57] Ayop, Osman, Mohamad Kamal A. Rahim, Noor Asniza Murad, and Noor Asmawati Samsuri. "Dual-resonant polarization-independent and wide-angle metamaterial absorber in X-band frequency." *Applied Physics A* 122, no. 4 (2016): 1-7.

[58] Li, Wei, Zhongyin Xiao, Xinyan Ling, Xiaoxia Zheng, Chuan Li, and Huanling Zou. "Design of a broadband, wide-angle metamaterial absorber base on hybrid

structure." *Journal of ElEctromagnEtic WavEs and applications* 32, no. 5 (2018): 651-660.

[59] Tuong, P. V., J. W. Park, J. Y. Rhee, K. W. Kim, W. H. Jang, H. Cheong, and Y. P. Lee. "Polarization-insensitive and polarization-controlled dual-band absorption in metamaterials." *Applied Physics Letters* 102, no. 8 (2013): 081122.

[60] Mousavi, Seyedmohammad A., Eric Plum, Jinhui Shi, and Nikolay I. Zheludev. "Coherent control of optical polarization effects in metamaterials." *Scientific reports* 5, no. 1 (2015): 1-11.

[61] Li, Yongqian, Lei Su, Chen Shou, Chunmeng Yu, Jinjun Deng, and Yu Fang. "Surface-enhanced molecular spectroscopy (SEMS) based on perfect-absorber metamaterials in the mid-infrared." *Scientific reports* 3, no. 1 (2013): 1-8.

[62] Cao, Tun, Chenwei Wei, Libang Mao, and Yang Li. "Extrinsic 2D chirality: giant circular conversion dichroism from a metal-dielectric-metal square array." *Scientific reports* 4, no. 1 (2014): 1-7.

[63] Chen, Hou-Tong. "Interference theory of metamaterial perfect absorbers." *Optics express* 20, no. 7 (2012): 7165-7172.

[64] Huang, Li, Dibakar Roy Chowdhury, Suchitra Ramani, Matthew T. Reiten, Sheng-Nian Luo, Abul K. Azad, Antoinette J. Taylor, and Hou-Tong Chen. "Impact of resonator geometry and its coupling with ground plane on ultrathin metamaterial perfect absorbers." *Applied Physics Letters* 101, no. 10 (2012): 101102.

[65] Rhee, J. Y., Y. J. Yoo, K. W. Kim, Y. J. Kim, and Y. P. Lee. "Metamaterial-based perfect absorbers." *Journal of Electromagnetic Waves and Applications* 28, no. 13 (2014): 1541-1580.

[66] Hu, ChengGang, Xiong Li, Qin Feng, Xu'Nan Chen, and XianGang Luo. "Investigation on the role of the dielectric loss in metamaterial absorber." *Optics Express* 18, no. 7 (2010): 6598-6603.

[67] Wanghuang, Tenglong, Weijian Chen, Yongjun Huang, and Guangjun Wen. "Analysis of metamaterial absorber in normal and oblique incidence by using interference theory." *Aip Advances* 3, no. 10 (2013): 102118.

[68] Tuong, P. V., J. W. Park, V. D. Lam, K. W. Kim, H. Cheong, W. H. Jang, and Y. P. Lee. "Simplified perfect absorber structure." *Computational materials science* 61 (2012): 243-247.

[69] Ding, Fei, Yanxia Cui, Xiaochen Ge, Yi Jin, and Sailing He. "Ultra-broadband microwave metamaterial absorber." *Applied physics letters* 100, no. 10 (2012): 103506.

[70] Sun, Jingbo, Lingyun Liu, Guoyan Dong, and Ji Zhou. "An extremely broad band metamaterial absorber based on destructive interference." *Optics Express* 19, no. 22 (2011): 21155-21162.

[71] Landy, N. I., C. M. Bingham, T. Tyler, N. Jokerst, D. R. Smith, and W. J. Padilla. "Design, theory, and measurement of a polarization-insensitive absorber for terahertz imaging." *physical review B* 79, no. 12 (2009): 125104.

[72] Shen, Xiaopeng, Tie Jun Cui, Junming Zhao, Hui Feng Ma, Wei Xiang Jiang, and Hui Li. "Polarization-independent wide-angle triple-band metamaterial absorber." *Optics express* 19, no. 10 (2011): 9401-9407.

[73] Ye, Yu Qian, Yi Jin, and Sailing He. "Omnidirectional, polarization-insensitive and broadband thin absorber in the terahertz regime." *JOSA B* 27, no. 3 (2010): 498-504.

[74] Ayop, Osman, Mohamad Kamal Abd Rahim, Noor Asniza Murad, Noor Asmawati Samsuri, and Raimi Dewan. "Triple band circular ring-shaped metamaterial absorber for x-band applications." *Progress In Electromagnetics Research M* 39 (2014): 65-75.

[75] Yao, Gang, Furi Ling, Jin Yue, Chunya Luo, Jie Ji, and Jianquan Yao. "Dual-band tunable perfect metamaterial absorber in the THz range." *Optics express* 24, no. 2 (2016): 1518-1527.

[76] Ma, Zhaofeng, and Fei Ding. "Ultra-broadband metamaterial absorber in Terahertz regime." In *2012 Asia Communications and Photonics Conference (ACP)*, pp. 1-3. IEEE, 2012.

[77] Cui, Yanxia, Kin Hung Fung, Jun Xu, Hyungjin Ma, Yi Jin, Sailing He, and Nicholas X. Fang. "Ultrabroadband light absorption by a sawtooth anisotropic metamaterial slab." *Nano letters* 12, no. 3 (2012): 1443-1447.

[78] Huang, Li, Dibakar Roy Chowdhury, Suchitra Ramani, Matthew T. Reiten, Sheng-Nian Luo, Antoinette J. Taylor, and Hou-Tong Chen. "Experimental demonstration of terahertz metamaterial absorbers with a broad and flat high absorption band." *Optics letters* 37, no. 2 (2012): 154-156.

[79] Munaga, Praneeth, Saptarshi Ghosh, Somak Bhattacharyya, and Kumar Vaibhav Srivastava. "A fractal-based compact broadband polarization insensitive metamaterial absorber using lumped resistors." *Microwave and Optical Technology Letters* 58, no. 2 (2016): 343-347.

[80] Shi, Yan, Yuan Chang Li, Tong Hao, Long Li, and Chang-Hong Liang. "A design of ultra-broadband metamaterial absorber." *Waves in random and complex media* 27, no. 2 (2017): 381-391.

[81] Ghobadi, Amir, Hodjat Hajian, Murat Gokbayrak, Sina Abedini Dereshgi, Ahmet Toprak, Bayram Butun, and Ekmel Ozbay. "Visible light nearly perfect absorber: an optimum unit cell arrangement for near absolute polarization insensitivity." *Optics express* 25, no. 22 (2017): 27624-27634.

[82] Shen, Guoqing, Ming Zhang, Yanping Ji, Wanxia Huang, Honglin Yu, and Jianping Shi. "Broadband terahertz metamaterial absorber based on simple multi-ring structures." *Aip Advances* 8, no. 7 (2018): 075206.

[83] Jiaming Hao, Jing Wang, XianLiang Liu, Willie J. Padilla, Lei Zhou, and Min Qiu. "High performance optical absorber based on a plasmonic metamaterial." *Applied Physics Letters* 96, no. 25(2010): 251104.

[84] Xiong, Han, Qi Shen, and Tian-Bin Long. "Broadband terahertz absorber based on Dirac semimetal with tunable working bandwidth." *Results in Physics* 19 (2020): 103683.

[85] Zhang, Ting, Sen Yang, and XinYing Yu. "Tunable multiple broadband terahertz perfect absorber based on vanadium dioxide." *Optics Communications* 501 (2021): 127358.

[86] Wang, Zongli, Xin Wang, Junlin Wang, Huizhong Pang, Suyalatu Liu, and Huqiang Tian. "Independently tunable dual-broadband terahertz absorber based on two-layer graphene metamaterial." *Optik* 247 (2021): 167958.

[87] Appasani, Bhargav. "A hybrid terahertz metamaterial sensor using a hexagonal ring resonator with bio-medical applications." *Plasmonics* 17, no. 2 (2022): 519-524.

[88] Gao, Zhiqiang, Qi Fan, Xiaoxia Tian, Cuilian Xu, Zhen Meng, Sining Huang, Tong Xiao, and Changhui Tian. "An optically transparent broadband metamaterial absorber for radar-infrared bi-stealth." *Optical Materials* 112 (2021): 110793.

[89] Lan, Feng, Zi-Fan Meng, Jiu-Fu Ruan, Rui-Zhi Zou, and Sheng-Wei Ji. "All-dielectric water-based metamaterial absorber in terahertz domain." *Optical Materials* 121 (2021): 111572.

[90] Liu, Wen, Jinping Tian, Rongcao Yang, and Weihua Pei. "Design of a type of broadband metamaterial absorber based on metal and graphene." *Current Applied Physics* 31 (2021): 122-131.

[91] Luo, Zhiyou, Shijun Ji, Ji Zhao, Han Wu, and Handa Dai. "A multiband metamaterial absorber for GHz and THz simultaneously." *Results in Physics* 30 (2021): 104893.

[92] Luo, Hao, Hai Liu, Cong Chen, Yue Feng, Peng Gao, Zi Yan Ren, and Yu Jia Qiao. "Dual-broadband terahertz absorber based on phase transition characteristics of VO2." *Results in Physics* 34 (2022): 105270.

[93] Nie, Pengcheng, Dongying Zhu, Zijian Cui, Fangfang Qu, Lei Lin, and Yue Wang. "Sensitive detection of chlorpyrifos pesticide using an all-dielectric broadband terahertz metamaterial absorber." *Sensors and Actuators B: Chemical* 307 (2020): 127642.

[94] Zhou, Jiangfeng, Eleftherios N. Economon, Thomas Koschny, and Costas M. Soukoulis. "Unifying approach to left-handed material design." *Optics letters* 31, no. 24 (2006): 3620-3622.

[95] Zhou, Jiangfeng, Lei Zhang, Gary Tuttle, Thomas Koschny, and Costas M. Soukoulis. "Negative index materials using simple short wire pairs." *Physical Review B* 73, no. 4 (2006): 041101.

[96] Lam, V. D., J. B. Kim, S. J. Lee, Y. P. Lee, and J. Y. Rhee. "Dependence of the magnetic-resonance frequency on the cut-wire width of cut-wire pair medium." *Optics Express* 15, no. 25 (2007): 16651-16656.

[97] Lam, V., Kim, J., Tung, N., Lee, S., Lee, Y. and Rhee, J.. "Dependence of the distance between cut-wire-pair layers on resonance frequencies." *Optics Express* 2008, 16(8), p.5934.

[98] Pang, Yong-qiang, Yong-jiang Zhou, and Jun Wang. "Equivalent circuit method analysis of the influence of frequency selective surface resistance on the frequency response of metamaterial absorbers." *Journal of Applied Physics* 110, no. 2 (2011): 023704.

[99] Yan-Zhao, Wang, Xu He-Xiu, Wang Chao-Hui, Wang Ming-Zhao, and Wang Shao-Jie. "Research progress of electromagnetic metamaterial absorbers." *Acta Physica Sinica* 69, no. 13 (2020).

[100] Rozanov, Konstantin N. "Ultimate thickness to bandwidth ratio of radar absorbers." *IEEE Transactions on Antennas and Propagation* 48, no. 8 (2000): 1230-1234.

[101] Acher, Olivier, and Sébastien Dubourg. "Generalization of Snoek's law to ferromagnetic films and composites." *Physical review B* 77, no. 10 (2008): 104440.
[102] Yihang Ding, Eleftherios Christos Statharas, Kui Yao, and Minghui Hong. "A broadband acoustic metamaterial with impedance matching layer of gradient index." *Applied Physics Letters*, 110, no. 24 (2017): 241903.
[103] Mou, Jinchao, and Zhongxiang Shen. "Broadband and thin magnetic absorber with non-Foster metasurface for admittance matching." *Scientific reports* 7, no. 1 (2017): 1-9.
[104] Mou, Jinchao, and Zhongxiang Shen. "Design and experimental demonstration of non-Foster active absorber." *IEEE Transactions on Antennas and Propagation* 65, no. 2 (2016): 696-704.
[105] Banadaki, Mohsen Dehghan, Abbas Ali Heidari, and Mansor Nakhkash. "A metamaterial absorber with a new compact unit cell." *IEEE Antennas and Wireless Propagation Letters* 17, no. 2 (2017): 205-208.
[106] Li, Wei, Jia Wei, Wei Wang, Dawei Hu, Yukun Li, and Jianguo Guan. "Ferrite-based metamaterial microwave absorber with absorption frequency magnetically tunable in a wide range." *Materials & Design* 110 (2016): 27-34.
[107] Cheng, Y. Z., Y. Nie, and R. Z. Gong. "Design of a thin wide-band absorber based on metamaterials and resistance frequency selective surface." *Acta Phys Sin* 61 (2012): 134101.
[108] Li, Wangchang, Qing Liu, Liwei Wang, Zuzhi Zhou, Jingwu Zheng, Yao Ying, Liang Qiao, Jing Yu, Xiaojing Qiao, and Shenglei Che. "Low frequency and broadband metamaterial absorber with cross arrays and a flaked iron powder magnetic composite." *AIP Advances* 8, no. 1 (2018): 015318.
[109] Zhang, Cheng, Qiang Cheng, Jin Yang, Jie Zhao, and Tie Jun Cui. "Broadband metamaterial for optical transparency and microwave absorption." *Applied Physics Letters* 110, no. 14 (2017): 143511.
[110] Sun, J. U. N., K. E. Chen, Guowen Ding, Wenlong Guo, Junming Zhao, Yijun Feng, and Tian Jiang. "Achieving directive radiation and broadband microwave absorption by an anisotropic metasurface." *IEEE Access* 7 (2019): 93919-93926.
[111] Peng, Lin, Xiao-feng Li, Xi Gao, Xing Jiang, and Si-min Li. "Methodology for the design of a multi-functional device with switchable absorption and polarization conversion modes by graphene and metallic metasurfaces." *Optical Materials Express* 9, no. 2 (2019): 687-705.
[112] Chen, Wenjie, Rui Chen, Yi Zhou, and Yungui Ma. "A switchable metasurface between meta-lens and absorber." *IEEE Photonics Technology Letters* 31, no. 14 (2019): 1187-1190.
[113] Wang, Jia Fu, Shao Bo Qu, Zhuo Xu, Zhen Tang Fu, Hua Ma, and Yi Ming Yang. "A broad-band three-dimensional isotropic left-handed metamaterial." *Journal of Physics D: Applied Physics* 42, no. 15 (2009): 155413.
[114] Lei, Lu, Qu Shao-Bo, Shi Hong-Yu, Zhang An-Xue, Zhang Jie-Que, and Ma Hua. "A miniaturized low-frequency polarization-insensitive metamaterial absorber based on broadside-coupled spiral structures." *ACTA PHYSICA SINICA* 62, no. 15 (2013).

[115] Xu, He-Xiu, Guang-Ming Wang, Zui Tao, and Tie Jun Cui. "High-directivity emissions with flexible beam numbers and beam directions using gradient-refractive-index fractal metamaterial." *Scientific reports* 4, no. 1 (2014): 1-10.

[116] Xu, He-Xiu, Guang-Ming Wang, Kai Ma, and Tie Jun Cui. "Superscatterer illusions without using complementary media." *Advanced Optical Materials* 2, no. 6 (2014): 572-580.

[117] Yi-Jun, Xiong, Wang Yan, Wang Qiang, Wang Chun-Qi, Huang Xiao-Zhong, Zhang Fen, and Zhou Ding. "Structural broadband absorbing metamaterial based on three-dimensional printing technology." *Acta Physica Sinica* 67, no. 8 (2018).

[118] Lim, Daecheon, Sungwook Yu, and Sungjoon Lim. "Miniaturized metamaterial absorber using three-dimensional printed stair-like Jerusalem cross." *IEEE Access* 6 (2018): 43654-43659.

[119] Jiang, Wei, Leilei Yan, Hua Ma, Ya Fan, Jiafu Wang, Mingde Feng, and Shaobo Qu. "Electromagnetic wave absorption and compressive behavior of a three-dimensional metamaterial absorber based on 3D printed honeycomb." *Scientific reports* 8, no. 1 (2018): 1-7.

[120] Xie, Jianwen, Sakib Quader, Fajun Xiao, Chong He, Xianling Liang, Junping Geng, Ronghong Jin, Weiren Zhu, and Ivan D. Rukhlenko. "Truly all-dielectric ultrabroadband metamaterial absorber: Water-based and ground-free." *IEEE Antennas and Wireless Propagation Letters* 18, no. 3 (2019): 536-540.

[121] Rashid, Amir K., Bo Li, and Zhongxiang Shen. "An overview of three-dimensional frequency-selective structures." *IEEE Antennas and Propagation Magazine* 56, no. 3 (2014): 43-67.

[122] Yu, Yufeng, Zhongxiang Shen, Tianwei Deng, and Guoqing Luo. "3-D frequency-selective rasorber with wide upper absorption band." *IEEE Transactions on Antennas and Propagation* 65, no. 8 (2017): 4363-4367.

[123] Li, Wei, Tianlong Wu, Wei Wang, Jianguo Guan, and Pengcheng Zhai. "Integrating non-planar metamaterials with magnetic absorbing materials to yield ultra-broadband microwave hybrid absorbers." *Applied Physics Letters* 104, no. 2 (2014): 022903.

[124] Li, Min, Lian Shen, Liqiao Jing, Su Xu, Bin Zheng, Xiao Lin, Yihao Yang, Zuojia Wang, and Hongsheng Chen. "Origami metawall: Mechanically controlled absorption and deflection of light." *Advanced science* 6, no. 23 (2019): 1901434.

Chapter 4

Microwave Metamaterial Sensors: Design Methodologies and Challenges

Aprajita Mishra, Divya Prakash[*] and Nisha Gupta
Department of Electronics and Communication Engineering,
Birla Institute of Technology, Mesra, Ranchi, India

Abstract

Apart from some of the useful applications of metamaterials like Superlenses, Cloaking, Perfect tunneling of electromagnetic waves, another exciting application of metamaterials is Metamaterial Sensors. The applications of the metamaterials are derived from their exotic properties such as negative permittivity and permeability. Split ring resonators (SRRs) having narrow absorption bandwidth, are planar structures that become very handy in the realization of Metamaterial sensors. These sensors are desirous in the sense that they offer an alternate sensing mechanism, are small, low cost, and easy to fabricate, thus finding suitability in the design of sensors to characterize materials depending upon their dielectric properties, to sense physical and chemical properties of materials. The chapter explores various configurations of metamaterial sensors. Each application has its specific requirement like the suitable placement of the Material Under Test (MUT) to achieve high sensitivity, which becomes an important design constraint. For addressing the intended sensor functionality various topologies of the SRRs and metamaterial structures are employed, which is elaborated in the chapter. Further, the method of excitation of the resonators is discussed, as it is a part of the sensor design process and determines the ease of operability. The important sensor parameters like

[*] Corresponding Author's Email: divyabit18@yahoo.com.

In: Metamaterials for Microwave and Terahertz Applications
Editors: B. Appasani, O. Prakash Acharya, A. Vidyakant Jha et al.
ISBN: 979-8-88697-330-3

sensitivity, selectivity, and quality factor; the ease of fabrication and operation, compactness, and energy consumption invariably depend upon the selected sensor configuration.

Keywords: metamaterials, metamaterial absorbers, microwave sensors, split-ring resonators

1. Introduction

Sensors have become an integral part of our everyday life. We find sensors embedded in numerous electronic gadgets of everyday use and as standalone devices to sense various parameters such as temperature, pressure, humidity, proximity, light intensity, touch, density, etc. Demand for sensors has grown in many folds with the automatization of industrial processes. With the advent of Artificial Intelligence in industry 4.0, the role of sensors has become even more critical as they provide first-hand information on which important decisions are taken. Among various sensors available for use, microwave sensors are gaining popularity due to their low fabrication cost, versatility, high sensitivity, robustness, and easy measurement procedure. These sensors can perform efficiently in different environments and find their suitability in various industrial, biomedical, and environmental applications. The two vital parameters of the sensor, the sensitivity, and resolution, can further be enhanced in a microwave sensor by embedding the concept of metamaterials.

In this chapter, we discuss sensors that operate by the manipulation of electromagnetic waves through the artificially engineered materials known as Metamaterials. Metamaterials are called artificial because they are geometrically or physically engineered rather than through any chemical synthesis. Metamaterials exhibit some unique properties such as negative permittivity and permeability, backward wave propagation, and negative refractive index [1-4]; utilizing these exotic properties, sensors are designed. These sensors provide an alternate sensing mechanism, are small, low cost, and can be fabricated easily with the available technologies.

Metamaterial-inspired sensors are realized with different metamaterial elements and structures. Surface Integrated Waveguide (SIW) Cavity resonators loaded with metamaterial patterns such as CSRR and interdigital capacitors have been used in research works [5-6], to build sensors, where the cavity itself is used to put the Material Under Test (MUT). Sensors designed on the phenomenon of cavity mode resonance (CMR); having a metasurface for achieving multiple reflections in the cavity, creating a resonance; have been illustrated in [7-8], where a narrow band absorption is achieved when the frequency of incoming waves match the resonant frequency of the cavity. The sensing parameter is the change in the resonant frequency that occurs as the cavity is exposed to the MUT. Composite right/left-handed transmission lines (CRLH-TL) are metamaterial elements that are also used in the design of sensors [9-12]. Here, the transmission line becomes the sensing element and its phase response varies when exposed to different MUTs. Sensors built from metamaterial transmission lines are generally planar devices where the CRLH-TL is constructed using a microstrip transmission line as a host medium. A microstrip transmission line is also used to couple electromagnetic energy to metamaterial structures like split-ring resonators of different shapes to build sensors [13-14]. A shift in the resonant frequency occurs when the MUTs are placed near the resonators. When the resonators are loaded with the MUTs, the reflections and transmissions through the microstrip line are used to sense materials having different dielectric properties and physical attributes like temperature, pressure, and density. Another exciting and popular application of metamaterials is the 'Absorber'. Metamaterial-based absorbers are planar structures built on dielectric materials that neither transmit nor reflect electromagnetic waves at a particular frequency, called the resonant frequency. The impedance of the free space matches that of the absorber surface at the resonant frequency, minimizing reflections, and further, the ohmic and dielectric losses prevent any wave transmission. The confinement of electromagnetic energy within the absorber at the resonant frequency is exploited to build sensors. The absorber is composed of a resonating structure built over a substrate, and the bottom-most layer of the absorber is a metallic ground plane that further prevents any transmission of electromagnetic energy, having a width greater than the skin depth. Metamaterial absorbers are broadly classified into broadband and narrowband absorbers. The broadband metamaterial absorbers find application in stealth technology, anechoic chambers, and Electromagnetic Interference (EMI) reduction, to name a few [15]. The metamaterial absorber-based sensors are primarily built of narrowband absorbers, as narrowband absorption is needed for high sensitivity and selectivity.

In recent years the popular application of metamaterial-based sensors has been the dielectric characterization of materials, measuring the concentration of chemical fluids, adulteration detection; sensing physical parameters like temperature, pressure, moisture, and density [16-17]; sensing biological tissues [18-19], blood glucose level monitoring [20], etc. Once an application has been decided, the next step is selecting an appropriate sensor configuration. The sensor is configured to achieve high sensitivity and selectivity with small size, low cost, and low power requirements. The topology of the resonators used, the placement of the material under test (MUT), and the sensor's excitation method ultimately decides the sensor configuration. Split ring resonators (SRRs), the common and basic building blocks of metamaterial sensors, can be used in various topologies to enhance interaction with the material under test (MUT) and achieve high sensitivity and selectivity. Generally, when it comes to metamaterial-inspired sensors, the sensing parameter is mainly the shift in the resonant frequency. Therefore, sensitivity is defined here as the amount of the resonant frequency shift upon the change in the MUT attributes. An important design consideration is placing the MUT at an appropriate place to maximize its interaction with the resonators and increasing the sensitivity by its placement around areas of high electric field concentration. The placement of the MUT should be such that it also makes the sensing operation effortless, consuming the least amount of time. Once a sensor configuration has been decided, the next step is simulating the sensor design and optimizing the design to achieve targeted sensor parameters. While simulation, it is to be remembered that the simulation setup should, as closely as possible, mimic the experimental environment of the fabricated sensor in operation. The errors in the sensor measurements arise mainly due to the difference in simulation and experimental environment, fabrication anomalies, and calibration errors. These errors should be tried and kept at a minimum to achieve the desired sensor performance. The final step is the fabrication of the optimized design. It is the quality of fabrication that impacts the sensor performance. Therefore, it must be ascertained that the components used, i.e., the substrates, connectors, etchants, etc., are authentic, masks prepared for the photolithographic process are meticulous, and the standard procedures are carried out for any fabrication process.

2. Sensing Principle

In general, the fundamental working principle of these sensors is the interaction of the materials under test (MUT) with the confined electromagnetic waves in the metamaterial structures at resonant frequencies. The resonance frequency is determined by the metamaterial structure's dimension and topology. At the resonance, electric fields established near the resonator make the region sensitive to dielectric changes. The MUT, placed near the resonator, causes a shift in the resonance frequency of the metamaterial structure. The greater the shift, the higher the sensitivity of the sensor. The shift in resonant frequency occurs because of effective variation of permittivity of the analyte medium, change of over-layer thickness, the variation of refractive index (RI), or any other physical parameter such as temperature, density, moisture of the MUT that alters the dielectric property of the substrate or the region in proximity of the resonator [21]. The resonance frequency of the unloaded resonator is given by equation (1).

$$f_r = \frac{1}{2\pi\sqrt{L_u C_u}} \quad (1)$$

where L_u and C_u are the inductance and capacitance of the unloaded resonator. The inductance in the equation arises from the current in the metal ring of the resonator, and the capacitance arises from the gap or split in the resonator rings and the capacitance between the metal portion of the rings and the ground plane. Therefore, any change in the resonator rings' dimension, the substrate width, or any change in the permittivity of the medium surrounding the resonator causes the resonant frequency to shift. The resonance frequency of the loaded resonator is given by equation (2).

$$f_{rL} = \frac{1}{2\pi\sqrt{L_u (C_u + C_L)}} \quad (2)$$

where C_L comes in the equation due to the loading sensor with MUT. A narrow absorption bandwidth is required in the case of sensors, as with the broad absorption bandwidth, it is difficult to discriminate between the resonant frequencies arising from the interaction with the MUTs. This is the sensor's selectivity, which is enhanced with narrow absorption bandwidth or the high-quality factor. The prime attribute considered while designing any sensor is its sensitivity. In the case of the sensors designed using metamaterial absorbers, greater absorption of the incident electromagnetic waves leads to greater confinement of electromagnetic energy around the resonators; in other words, it leads to stronger resonance. Stronger resonances lead to higher sensitivities causing even a slight change in the effective permittivity to change the resonance frequency.

3. Sensor Configurations and Design Challenges

The sensor configuration is decided by the application, topology of the resonators used, the placement of material under test (MUT), and the excitation method. Selecting the right resonator topology is essential as the sensor sensitivity is decided by the concentration of electric fields and the interaction of the MUT with the resonators. The sensor may be composed of an isolated unit cell or an array of unit cells [17, 22-23]. Even a unit cell can have multiple resonators with different dimensions. Multiple resonators in a unit cell help achieve multiband resonance, increasing the accuracy and reliability of the sensor [24]. The split ring resonators (SRR) in a unit cell can be edge coupled or broadside coupled, having circular or rectangular shapes in general. The edge-coupled resonators lie in the same plane, while the broadside coupled resonators are realized on substrates on different planes. There are many research papers where the MUT is placed between the broadside coupled resonators, the rationale being increasing the interaction of the MUT with the resonators [25].

In the case of single or edge coupled resonators, the MUT is placed near the split region, where the concentration of electric fields is maximum. Complimentary Split Ring Resonators (CSRR) provide the electric field concentration throughout the etched ring, thereby increasing the interaction area with the MUT and thus the sensitivity [21, 26]. MUTs with larger dimensions can be used with CSRR topology.

Placement of the MUT in proximity to the SRRs to achieve greater interaction and hence greater sensitivity is a design challenge. Applications focused on finding the concentration of liquids and their dielectric characterization, adulteration of liquid fuels, used specially designed channels for the fluid MUT to flow and interact with the resonators. In [24] a sensor is proposed where a microfluidic channel is built beneath the substrate for the liquid MUT, whose concentration is to be sensed. A sensor to find ethanol concentration in a binary mixture with water is discussed in [27]. Here, a hole is drilled in the substrate, and a tube is inserted to pass the liquid mixture.

An alternate method to sense liquid substances is proposed in [28], where a filter paper is placed on top of a coplanar waveguide structure, and a measured amount of liquid MUT is dropped on the filter paper. A similar approach is also suggested in [29], where the filter paper carrying the MUT is placed between the resonators. The placement of solid MUTs is simpler as compared to liquid MUTs. The solid MUT can be placed above the resonators, like in [30], to determine the dielectric characteristics of the MUT; it is placed above the S-shaped resonator, Figure 1.

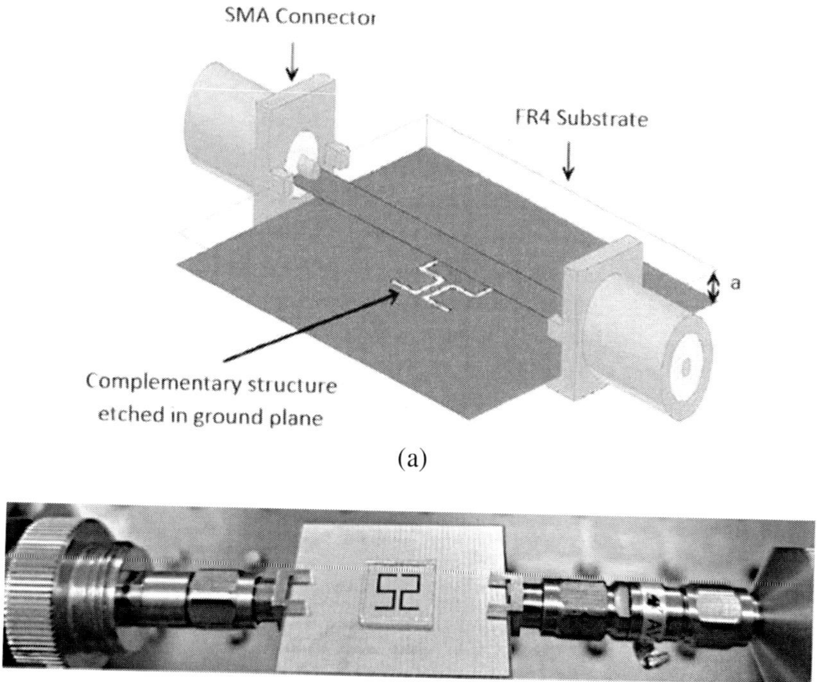

(a)

(b)

Figure 1. (a) Simulation structure of the S-shaped resonator with microstrip line. (b) Experimental setup of S shaped resonator (copyright from [30]).

The solid MUTs can even be placed in between the resonators [29]. A metamaterial-inspired sensor needs to be excited by an electromagnetic wave along with the MUT to observe its reflection and transmission behavior. The sensor can be excited through a transmission line, by an antenna, or the entire sensor with the MUT can be placed within a waveguide or section of two waveguides for the excitation. In most of the research works reported so far, a Vector Network Analyzer (VNA) is used as a source of the electromagnetic wave, and the reflection and the transmission coefficients are measured. The type of excitation used decides the sensor configuration, ease of operation, and even functionality. When the sensor is excited through a transmission line, the reflection and the transmission coefficients can be observed easily on a VNA, but the sensor needs to be in proximity to the VNA; this makes remote sensing difficult. Besides, the SMA connectors are soldered to the transmission line, increasing the complexity and cost of the sensor. The advantage of using a transmission line as an excitation source is that the sensor can be easily integrated with other electronic circuitry, like the signal amplifiers, transmitters and receivers, and other signal conditioning circuits. In [20], a portable sensor is proposed for the non-invasive monitoring of blood glucose levels. Four hexagonal CSRRs are arranged in a honey cell design on the ground plane of an FR4 dielectric substrate and are coupled using a planar microstrip line to a RADAR board operating at 2.4-2.5 GHz. Honey cell design induces an electric field over an area large enough for a fixture of a fingertip, as shown in Figure 2. The signal transmitted through the sensor's transmission line is shown on the radar receiving port, which is analyzed using machine learning and signal processing techniques to identify various glucose levels. Another example of a sensor excited through a transmission line is [31], where multiple CSRRs are used for sensing dielectric substrates non-destructively, as shown in Figure 3.

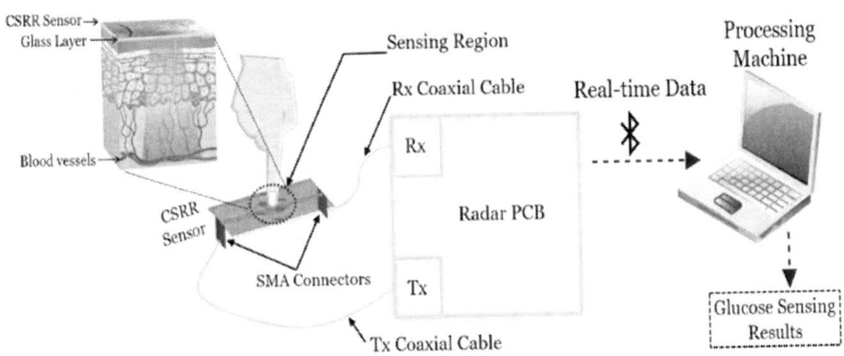

Figure 2. Experimental setup of blood glucose level monitoring system using honey cell design of hexagonal CSRR (copyright from [20]).

Excitation of the sensors through antennas makes remote sensing possible. A single antenna for both, as a wave source and as a receptor of the reflected signal, is placed in the far-field to the sensor [16, 32] or there can be separate antennas for transmission and reception placed in the far-field region [17]. Generally, an array of metamaterial unit cells is used in the sensor, where the excitation is accomplished through an antenna. An array of metamaterial unit cells in a sensor provides more gain as compared to the sensor having an isolated unit cell. Though there is an advantage of remote sensing through antenna excitation, there is also a greater scope of interference by the ambient environment while taking measurements. As connectors are not required, the design becomes simpler, and the portability of the sensor is achieved. As shown in Figure 4, a pair of antennas are used to excite an array of metamaterial unit cells of the sensor [17] and to capture the reflected signals from the array, the application being remote sensing of soil moisture content.

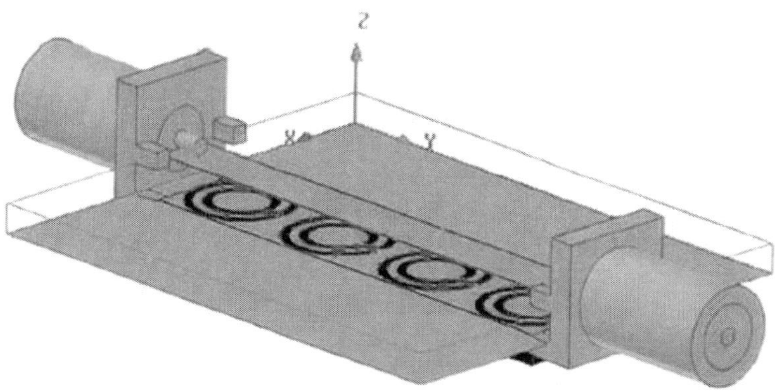

(a)

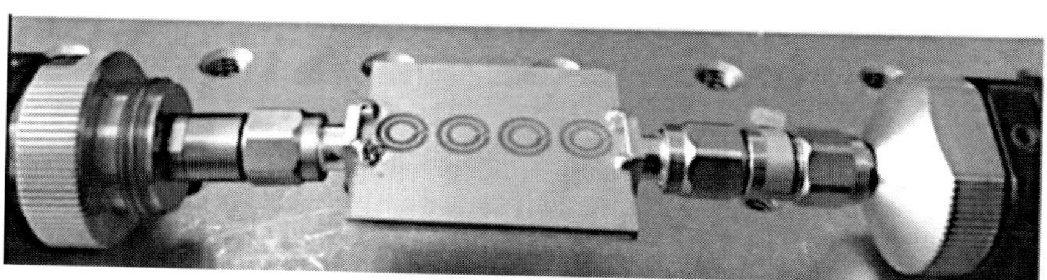

(b)

Figure 3. (a) Simulation setup of multiple CSRRs using transmission line excitation. (b) Experimental setup of multiple CSRRs using transmission line excitation (copyright from [31]).

In [16], a metamaterial absorber-based sensor is proposed that has two resonators, one of square shape and the second one ring-shaped, placed within the outer square resonator, with the splits aligned. The sensor is composed of 5×5-unit cells. The proposed design is tested for both absorber and sensor configurations depicting a resonance frequency between 6-8 GHz. The two resonators are used to increase the sensitivity, and the sensor can be utilized in various sensing applications such as pressure, temperature, density, and moisture sensing.

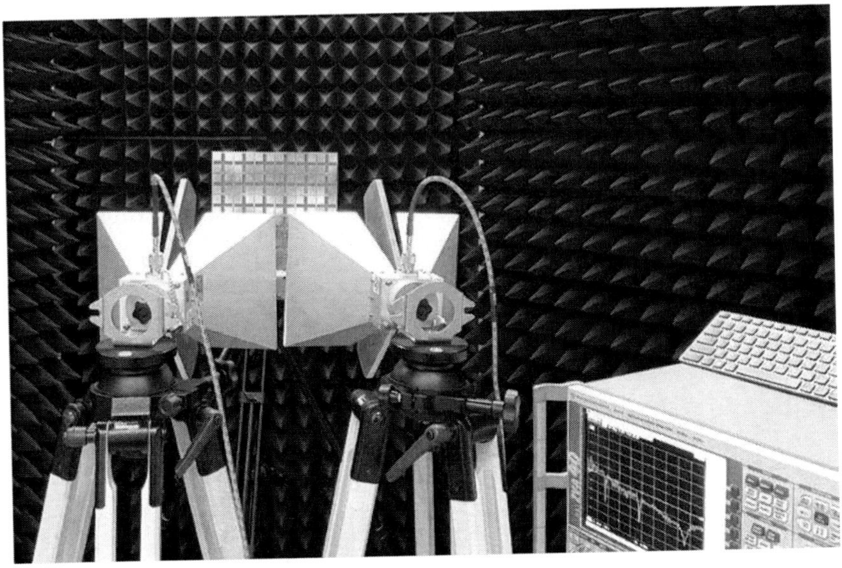

Figure 4. Experimental setup of sensor excitation using two antennas for soil moisture sensing (copyright from [17]).

The sensors can also be excited by placing them within a waveguide along with the MUT. Although the measurements utilize an accurate waveguide setup, the sensors' dimensions must match the waveguide's dimensions for the sensor to fit within and take measurements, this becomes a design constraint. Here, the signal originates from one of the ports of the VNA, and the reflected and transmitted signals are measured on it. The sensor and the MUT can be placed in the waveguide sample holder easily, but the placement of complex channels for the MUT is difficult in such a setup. Connectors are also not required in the waveguide setup, making the sensor design simpler, but connections with external electronic circuitry are also impossible. For applications involving concentration and humidity sensing, a double-sided topology of resonators is proposed in [29]. Here, sensors are designed to fit within the *S*-band and *X*-Band waveguide sample holders. The material whose concentration or humidity is to be sensed is placed between the resonators.

In [33], a metamaterial absorber-based sensor for sensing liquid chemicals is proposed. It consists of a swastika-shaped resonator on top, a copper plate on the backside, and an air gap to fill the chemical samples. The sensor configuration resonates at 10.12 GHz and 10.5 GHz frequencies. The sensor is placed within the waveguide along with the MUT for the excitation, and the reflection, transmission coefficients are observed on a VNA.

For non-destructive sensing of grain samples [7] proposes a sensor composed of a cavity and periodically arranged cross-resonators on a thin substrate. The sensor works on multiple reflections and interference that create resonance in the cavity where the MUT is placed, Figure 5. The resonance frequency shifts when the grains placed in the cavity are changed, owing to their dielectric properties. The sensor can also differentiate between different varieties of food grain, as each grain type placed in the cavity provides a unique resonant frequency. The experimental setup consists of two horn antennas, one to excite the sensor and the other is the detector antenna. The signals received from the detector antenna are analyzed on a VNA. Table 1 shows metamaterial sensing elements other than conventional split-ring resonators and their application.

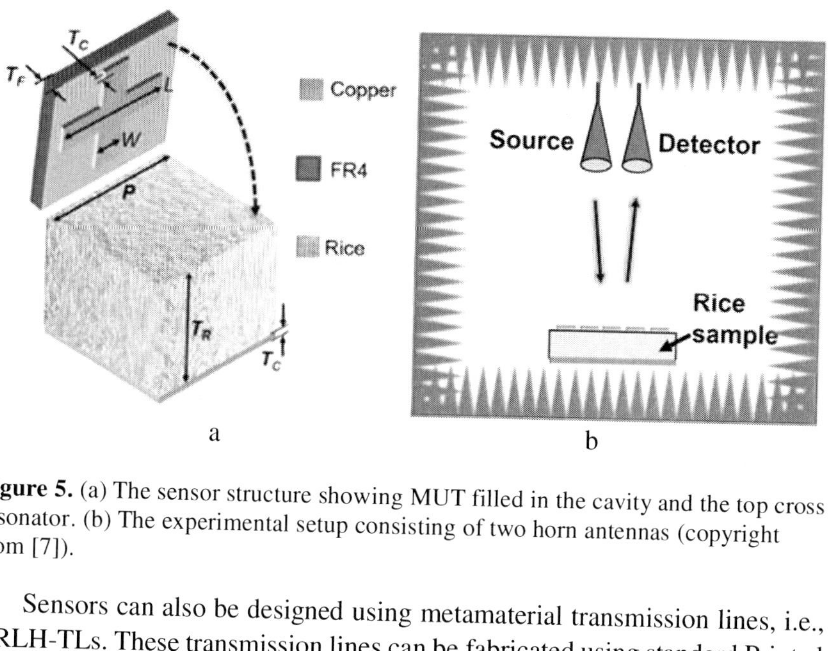

Figure 5. (a) The sensor structure showing MUT filled in the cavity and the top cross resonator. (b) The experimental setup consisting of two horn antennas (copyright from [7]).

Sensors can also be designed using metamaterial transmission lines, i.e., CRLH-TLs. These transmission lines can be fabricated using standard Printed Circuit Board technology and integrated with external electronic circuitry. For dielectric spectroscopy of chemical liquids [11] proposes a sensor composed of two CRLH-TLs, built on a right-handed transmission line host medium. A CSRR is also etched in the ground plane to make the structure a Left-Handed or double negative medium. Using polydimethylsiloxane (PDMS) fluidic channels, the MUT is placed in contact with the transmission line. The placement of the MUT, analogous to the change in dielectric constant, causes a phase variation sensed by the transmission line's output power. Sensors built from normal transmission lines provide larger sensing bandwidth with lesser sensitivity. In sensing applications, there is a trade-off between the bandwidth and sensitivity, but by using CRLH-TLs, optimum values of both the parameters can be attained [11]. Few more sensors using CRLH-TLs like Mass Flow Sensor, Differential Sensor, Wireless Strain, and Temperature Sensor, Level Sensor are elaborated in [12]. It has been concluded that with the possibility of scaling in size and frequency, these metamaterial sensors find numerous applications with slight modifications. Table 2 shows various sensor excitation methods along with the location of the MUT, the resonating structures, and the applications.

Table 1. Metamaterial sensing elements and their applications

Sr. No.	Reference	Sensing Element	Application
1	[5]	Metamaterial Inspired Surface Integrated Waveguide (SIW) cavity resonator loaded with interdigital capacitors.	Characterization of liquid samples based on permittivity.
2	[6]	Quarter Mode SIW resonator loaded with CSRRs.	Microfluidic chemical liquid characterization.
3	[17]	Metamaterial Perfect Absorber (MPA) using SIW cavity resonator.	Remote Sensing of Soil Moisture.
4	[7]	Cavity with periodically arranged cross-resonators.	Sensing grain quality and variety non-destructively
5	[8]	One dimensional metamaterial Cavity Mode Resonator.	High-Quality factor THz sensor.
6	[9]	Composite Right/Left-Handed Transmission Lines (CRLH-TLs)	Dielectric Spectroscopy of Chemical Liquids as an Interferometry Sensor.
7	[11]	Pair of 180-degree phase shifted Composite Right/Left-Handed Transmission Lines (CRLH-TLs)	Wide band interferometry sensor. Detection of chemical liquids, dielectric spectroscopy.
8	[47]	Composite Right/Left-Handed structure comprising interdigital capacitors.	Non-invasive monitoring of Blood Glucose.

Table 2. Various resonating structures with their excitation schemes, MUT placement, and applications

Sr. No.	Reference	Type of excitation	Placement of MUT	Resonating Structure	Application
1	[5]	Transmission line	The liquid sample is located at the centre hole through the cavity.	SIW cavity	Testing concentration of binary mixture of water and alcohol.
2	[13]	Transmission line	Channel of liquid samples is manually positioned across the SRR gap.	Split Ring Resonator	Complex permittivity characterization of different liquid samples.
3	[14]	Transmission Line	On the top of the copper track on the substrate.	Symmetrical Split Ring Resonator	Dielectric characterization of solid MUTs.
4	[37]	Waveguide	On the backside of the sensor structure.	G shaped resonator	Liquid chemicals detection.
5	[33]	Waveguide	Inside the air gap between resonator and backside metallic layer.	Swastika shaped resonator	Determination of the liquid chemicals depending on electrical characteristics.
6	[48]	Waveguide	Sensor layer is placed just behind the back side of the CMS structure in the waveguide.	Chiral Metamaterial sensor structure based on split ring resonator	Checking fuel adulteration.
7	[34]	Antenna	A microfluidic channel is etched on the top layer of substrate for placement of MUT.	Split-ring-cross resonators	Large scale ethanol sensor application.
8	[17]	Antenna	MUT is placed in two perpendicular slots etched in substrate.	Cavity Resonator	Soil moisture sensor.
9	[16]	Antenna	MUT is placed between the copper plate and the resonator layer.	Two split ring resonators.	Pressure, temperature, density, and moisture sensor.

4. Materials in Sensor Design

To achieve the desired sensor functionalities, various materials are employed. Choosing a suitable material is critical in designing channels and containers for holding MUTs. In [34], a wireless Ethanol Metamaterial Absorber based sensor is proposed where the liquid MUT is placed around the resonators. In this work, on the topmost layer, a microfluidic channel is etched on a Polydimethylsiloxane (PDMS) substrate, the middle layer has an inkjet-printed Split Ring Cross Resonator (SRCR) on a photo paper, and the bottom-most layer is a PDMS substrate that minimizes transmission, as shown in Figure 6.

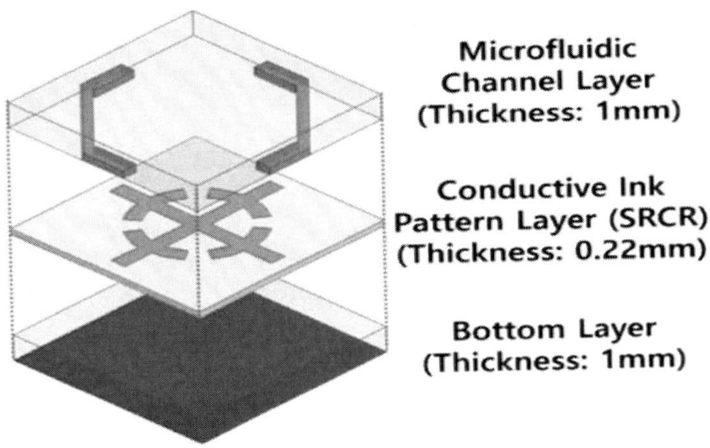

Figure 6. Layered structure of the sensor showing the inkjet printed resonator on a photopaper, the microfluidic channel and the PDMS bottom layer (copyright from [34]).

In [26], for dielectric characterization of fluids, a sensor inspired by Metamaterials is designed; where to bring the fluid MUT to the sensing area, a microfluidic PDMS channel is attached to the edge of the CSRR. Thin Kapton films are also used in sensor designs to seal the liquid waveguide sample holders, as shown in [35]. A different approach to deliver the liquid MUTs around the sensing region is adopted in [29], where a filter paper wetted with the (liquid under test) LUT is placed between the resonators for sensing concentrations. A similar technique is adopted in [28] for sweat monitoring non-invasively. Here, a filter paper absorbed with the sweat (LUT) is used as a superstrate and is placed over a CPW (coplanar waveguide) transmission line. Tunability of the resonance frequency is a desired attribute in metamaterial-based sensors. For achieving tunability, different electronic components and special materials are used. In [32], a tunable metamaterial absorber is proposed for sensing application. Here, a varactor diode is placed in a rectangular split ring resonator in one of the splits, and by changing the diode's reverse bias voltage, the resonator's desired resonance frequency is achieved. For attaining the tunability, a graphene layer is added over the substrate [36], as shown in Figure 7. By applying the external gate voltage, the surface conductivity of the graphene layer changes, which in turn changes the resonance frequency of the structure. Selection of the substrate with low permittivity is also essential in achieving greater sensitivity, as discussed in [24].

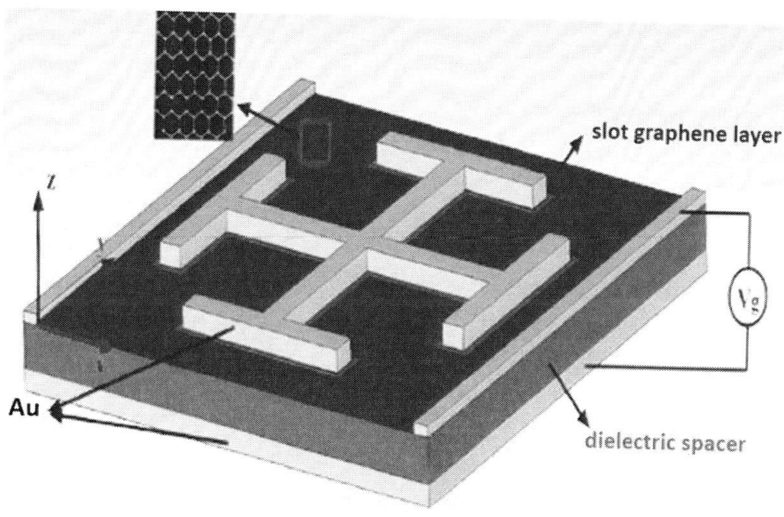

Figure 7. Prospective view of the sensor showing the graphene layer, resonator and voltage being applied to the graphene layer (copyright from [36]).

Metamaterial-inspired sensors comprise mainly of two parts: the substrate and the metal layer. Even though the properties of metamaterial-based structures depend mainly on their geometry, the material of the substrate and the metal layer have their importance both in terms of physical properties as well as fabrication. Dielectric substrates used can be distinguished broadly in two categories, i.e., Hard dielectrics like FR4 [37], Rogers RO3035 [27]; and soft dielectrics like Paper [38], various polymers including PDMS [39], and polymethylmethacrylate (PMMA) [40]. Hard dielectrics make the sensors durable, whereas soft dielectrics provide flexibility. The flexible sensors can be mounted over curved surfaces and are suitable for wearable applications. Other than the conventional methods of fabrication like Photolithography, Focused Ion beam (FIB), shadow mask lithography (SML), electron beam lithography (EBL), laser micro-lens array lithography, electroplating, and direct laser writing (DLW) [41], used for both hard and soft dielectrics; few other methods like inkjet printing [42], Direct drawing [43], etc. are employed for the fabrication of flexible metamaterial structures.

After selecting the sensor application, resonator topology, and the operating frequencies; the next step is the simulation of the design. While simulating the design, the boundary conditions are the most important aspect. A suitable boundary condition should be chosen to suit the experimental environment after the fabrication of the sensor. When a waveguide is used for exciting the sensor, Perfect Electric Conductor (PEC) boundary condition is chosen along the resonator axis that coincides with the walls of the waveguide, and open boundary conditions are chosen along the axis through which the wave propagates. The sensor can be an isolated unit cell or a periodic array of unit cells. Open boundary conditions are used when the sensor is an isolated metamaterial unit cell; otherwise, the unit cell boundary conditions are used when the sensor is an array of unit cells [44]. In the simulation, only the unit cell is simulated with periodic or unit cell boundary conditions, while such a sensor is fabricated having an array of unit cells.

Another challenging aspect in the design of sensors is the calibration. Calibration involves comparing the values obtained by the designed sensor with the known standard values. The calibration process is time-consuming as it requires a lot of data acquired experimentally to build a model which represents the sensor behavior. A robust model can be developed through the curve fitting method, machine learning (ML), or Principal Component Analysis (PCA) approach [24, 29, 45-46], which predicts the required parameters through the values obtained through the sensor. The main criterion for selecting a model is driven by the accuracy it offers.

Conclusion

In this chapter, the designs and all the steps involved in designing metamaterial-inspired sensors are elaborated. The concepts are elucidated to facilitate the readers to design low-cost sensors with a simple design and low power requirements. The topology of the resonators, and considerations for placement of the MUTs are explained for achieving high sensor sensitivity. Factors that affect the sensor accuracy and performance like the simulation environment, fabrication, and calibration are also discussed in detail. Various techniques for designing the sensors using different metamaterial elements line CRLH-TLs, SRRs, SIW resonators with embedded SRRs, Cavity Mode Resonators, microstrip lines, and metamaterial absorbers are presented. Metamaterial-based sensors are passive devices with some unique advantages and enable the possibility of remote sensing. These sensors can also be easily integrated with other electronic devices and circuitry. The scope of the addition of electronic lumped elements enhances their functionality. The contactless, non-destructive sensing using a minimal amount of sample to be tested is an added advantage. Owing to their simple structure, the sensors can be fabricated in bulk in less time and cost with the existing fabrication technologies. A wide range of operational frequencies, extending from the GHz to THz range, is achievable by simply scaling the sensor dimensions. However, in the THz range, the extremely small size of the sensors demands advanced fabrication techniques. The metamaterial-based sensors can potentially find applications in myriad industries, such as agriculture, healthcare, automobile, manufacturing, etc.

References

[1] Veselago, Victor Georgievich. "The electrodynamics of substances with simultaneously negative values of ε and μ." *Physics-Uspekhi* [*Physics-Successes*], 10, no. 4 (1968): 509-514.

[2] Pendry, John. "Metamaterials in the sunshine." *Nature Materials* 5, no. 8 (2006): 599-600.

[3] Pendry, John Brian. "Negative refraction makes a perfect lens." *Physical review letters* 85, no. 18 (2000): 3966.

[4] Smith, David R. "How to build a superlens." *Science* 308, no. 5721 (2005): 502-503.

[5] Yan, Sen, Juncheng Bao, Ilja Ocket, Bart Nauwelaers, and Guy AE Vandenbosch. "Metamaterial inspired miniaturized SIW resonator for sensor applications." *Sensors and Actuators A: Physical* 283 (2018): 313-316.

[6] Cai, Jing, Yong Jin Zhou, and Xin Mi Yang. "A metamaterials-loaded quarter mode SIW microfluidic sensor for microliter liquid characterization." *Journal of Electromagnetic Waves and Applications* 33, no. 3 (2019): 261-271.

[7] Zhang, Yin, Junming Zhao, Jie Cao, and Bo Mao. "Microwave metamaterial absorber for non-destructive sensing applications of grain." *Sensors* 18, no. 6 (2018): 1912.

[8] Liu, Zizheng, Luyao Wang, Ming Hua, Xiaoyu Liu, Fuyue Qian, Guangyin Xie, Yafei Ning, Yanpeng Shi, Xiaodong Wang, and Fuhua Yang. "High-Q metamaterials based on cavity mode resonance for THz sensing applications." *AIP Advances* 10, no. 7 (2020): 075014.

[9] Saghati, Ali Pourghorban, Jaskirat Singh Batra, Jun Kameoka, and Kamran Entesari. "A metamaterial-inspired wideband microwave interferometry sensor for dielectric spectroscopy of liquid chemicals." *IEEE Transactions on Microwave Theory and Techniques* 65, no. 7 (2017): 2558-2571.

[10] Schueler, Martin, Christian Mandel, Margarita Puentes, and Rolf Jakoby. "Metamaterial inspired microwave sensors." *IEEE Microwave Magazine* 13, no. 2 (2012): 57-68.

[11] Saghati, Ali Pourghorban, Jaskirat Singh Batra, Jun Kameoka, and Kamran Entesari. "A metamaterial-inspired miniaturized wide-band microwave interferometry sensor for liquid chemical detection." In *2016 IEEE MTT-S International Microwave Symposium (IMS),* pp. 1-3. IEEE, 2016.

[12] Mayani, Mahdieh Gholami, Francisco Javier Herraiz-Martínez, Javier Matanza Domingo, and Romano Giannetti. "Resonator-based microwave metamaterial sensors for instrumentation: Survey, classification, and performance comparison." *IEEE Transactions on Instrumentation and Measurement* 70 (2020): 1-14.

[13] Ma, Jialu, Jingchao Tang, Kaicheng Wang, Lianghao Guo, Yubin Gong, and Shaomeng Wang. "Complex permittivity characterization of liquid samples based on a split ring resonator (SRR)." *Sensors* 21, no. 10 (2021): 3385.

[14] Alahnomi, Rammah Ali, Zahriladha Zakaria, Eliyana Ruslan, Siti Rosmaniza Ab Rashid, and Amyrul Azuan Mohd Bahar. "High-Q sensor based on symmetrical split ring resonator with spurlines for solids material detection." *IEEE Sensors Journal* 17, no. 9 (2017): 2766-2775.

[15] Tirkey, Manish Mathew, and Nisha Gupta. "The quest for perfect electromagnetic absorber: a review." *International Journal of Microwave and Wireless Technologies* 11, no. 2 (2019): 151-167.

[16] Bakır, Mehmet, Muharrem Karaaslan, E. Unal, O. Akgol, and Cumali Sabah. "Microwave metamaterial absorber for sensing applications." *Opto-Electronics Review* 25, no. 4 (2017): 318-325.

[17] Amiri, Majid, Mehran Abolhasan, Negin Shariati, and Justin Lipman. "Soil moisture remote sensing using SIW cavity based metamaterial perfect absorber." *Scientific Reports* 11, no. 1 (2021): 1-17.

[18] Nejat, Mohamad, and Najmeh Nozhat. "Sensing and switching capabilities of a graphene-based perfect dual-band metamaterial absorber with analytical methods." *JOSA B* 37, no. 5 (2020): 1359-1366.

[19] Kumari, Ratnesh, Piyush N. Patel, and Rahul Yadav. "An ENG-inspired microwave sensor and functional technique for label-free detection of aspergillus Niger." *IEEE Sensors Journal* 18, no. 10 (2018): 3932-3939.

[20] Omer, Ala Eldin, George Shaker, Safieddin Safavi-Naeini, Hamid Kokabi, Georges Alquié, Frédérique Deshours, and Raed M. Shubair. "Low-cost portable microwave sensor for non-invasive monitoring of blood glucose level: Novel design utilizing a four-cell CSRR hexagonal configuration." *Scientific Reports* 10, no. 1 (2020): 1-20.

[21] Prakash, Divya, and Nisha Gupta. "Applications of metamaterial sensors: a review." *International Journal of Microwave and Wireless Technologies* 14, no. 1 (2022): 19-33.

[22] Chuma, Euclides Lourenço, Yuzo Iano, Sergio Barcelos, Luis Ernesto Ynoquio Herrera, Laez Barbosa da Fonseca Filho, and Rodolfo Cruz. "Surface crack detection in pipelines using CSRR microwave based sensor." *Progress In Electromagnetics Research* C 105 (2020): 11-21.

[23] Chuma, Euclides Lourenço, Yuzo Iano, Glauco Fontgalland, Leonardo Lorenzo Bravo Roger, and Hermes Loschi. "PCB-integrated non-destructive microwave sensor for liquid dielectric spectroscopy based on planar metamaterial resonator." *Sensors and Actuators A: Physical* 312 (2020): 112112.

[24] Zhou, Hong, Donglin Hu, Cheng Yang, Cong Chen, Junwang Ji, Ming Chen, Yu Chen, Ya Yang, and Xiaojing Mu. "Multi-band sensing for dielectric property of chemicals using metamaterial integrated microfluidic sensor." *Scientific reports* 8, no. 1 (2018): 1-11.

[25] Ekmekci, Evren, and Gonul Turhan-Sayan. "Multi-functional metamaterial sensor based on a broad-side coupled SRR topology with a multi-layer substrate." *Applied Physics A* 110, no. 1 (2013): 189-197.

[26] Ebrahimi, Amir, Withawat Withayachumnankul, Said Al-Sarawi, and Derek Abbott. "High-sensitivity metamaterial-inspired sensor for microfluidic dielectric characterization." *IEEE Sensors Journal* 14, no. 5 (2013): 1345-1351.

[27] Chuma, Euclides Lourenço, Yuzo Iano, Glauco Fontgalland, and Leonardo Lorenzo Bravo Roger. "Microwave sensor for liquid dielectric characterization based on metamaterial complementary split ring resonator." *IEEE Sensors Journal* 18, no. 24 (2018): 9978-9983.

[28] Riad, Michael MYR, and Angie R. Eldamak. "Coplanar waveguide based sensor using paper superstrate for non-invasive sweat monitoring." *IEEE Access* 8 (2020): 177757-177766.

[29] Ekmekci, Evren, Umut Kose, Alparslan Cinar, Ozgur Ertan, and Zeynep Ekmekci. "The use of metamaterial type double-sided resonator structures in humidity and concentration sensing applications." *Sensors and Actuators A: Physical* 297 (2019): 111559.

[30] Haq, Tanveerul, Cunjun Ruan, Xingyun Zhang, Shahid Ullah, Ayesha Kosar Fahad, and Wenlong He. "Extremely sensitive microwave sensor for evaluation of dielectric characteristics of low-permittivity materials." *Sensors* 20, no. 7 (2020): 1916.

[31] Haq, Tanveer ul, Cunjun Ruan, Xingyun Zhang, and Shahid Ullah. "Complementary metamaterial sensor for nondestructive evaluation of dielectric substrates." *Sensors* 19, no. 9 (2019): 2100.

[32] Bakır, Mehmet, Muharrem Karaaslan, Furkan Dinçer, K. Delihacioglu, and Cumali Sabah. "Tunable perfect metamaterial absorber and sensor applications." *Journal of Materials Science: Materials in Electronics* 27, no. 11 (2016): 12091-12099.

[33] Abdulkarim, Yadgar I., Lianwen Deng, Olcay Altıntaş, Emin Ünal, and Muharrem Karaaslan. "Metamaterial absorber sensor design by incorporating swastika shaped resonator to determination of the liquid chemicals depending on electrical

characteristics." *Physica E: Low-dimensional systems and Nanostructures* 114 (2019): 113593.

[34] Kim, Hyung Ki, Dongju Lee, and Sungjoon Lim. "A fluidically tunable metasurface absorber for flexible large-scale wireless ethanol sensor applications." *Sensors* 16, no. 8 (2016): 1246.

[35] Bakır, Mehmet, Şekip Dalgaç, Muharrem Karaaslan, Faruk Karadağ, Oğuzhan Akgol, Emin Unal, Tolga Depci, and Cumali Sabah. "A comprehensive study on fuel adulteration sensing by using triple ring resonator type metamaterial." *Journal of the Electrochemical Society* 166, no. 12 (2019): B1044.

[36] Huang, Hailong, Hui Xia, Wenke Xie, Zhibo Guo, Hongjian Li, and Ding Xie. "Design of broadband graphene-metamaterial absorbers for permittivity sensing at mid-infrared regions." *Scientific reports* 8, no. 1 (2018): 1-10.

[37] Abdulkarim, Y. I., Deng, L., Luo, H., Huang, S., Karaaslan, M., Altıntaş, O., Bakır, M., Muhammadsharif, F. F., Awl, H. N., Sabah, C., & Lateef Al-badri, K. S. "Design and study of a metamaterial based sensor for the application of liquid chemicals detection." *Journal of Materials Research and Technology* 9, no. 5 (2020): 10291-10304.

[38] Yoo, Minyeong, Hyung Ki Kim, Sangkil Kim, Manos Tentzeris, and Sungjoon Lim. "Silver nanoparticle-based inkjet-printed metamaterial absorber on flexible paper." *IEEE Antennas and Wireless Propagation Letters* 14 (2015): 1718-1721.

[39] Kim, Kyeongseob, Dongju Lee, Seunghyun Eom, and Sungjoon Lim. "Stretchable metamaterial absorber using liquid metal-filled polydimethylsiloxane (PDMS)." *Sensors* 16, no. 4 (2016): 521.

[40] Shi, Zaiying, Lixin Song, and Tao Zhang. "Optical and electrical characterization of pure PMMA for terahertz wide-band metamaterial absorbers." *Journal of Infrared, Millimeter, and Terahertz Waves* 40, no. 1 (2019): 80-91.

[41] Walia, Sumeet, Charan M. Shah, Philipp Gutruf, Hussein Nili, Dibakar Roy Chowdhury, Withawat Withayachumnankul, Madhu Bhaskaran, and Sharath Sriram. "Flexible metasurfaces and metamaterials: A review of materials and fabrication processes at micro-and nano-scales." *Applied Physics Reviews* 2, no. 1 (2015): 011303.

[42] Tirkey, Manish Mathew, and Nisha Gupta. "Broadband polarization-insensitive inkjet-printed conformal metamaterial absorber." *IEEE Transactions on Electromagnetic Compatibility* 63, no. 6 (2021): 1829-1836.

[43] Wang, Zhongyang, Xueyan Fu, Zidong Zhang, Yuliang Jiang, Moaz Waqar, Peitao Xie, Ke Bi, Yao Liu, Xiaowei Yin, and Runhua Fan. "based metasurface: Turning waste-paper into a solution for electromagnetic pollution." *Journal of Cleaner Production* 234 (2019): 588-596.

[44] Turkmen, Oznur, Evren Ekmekci, and Gonul Turhan-Sayan. "Effects of using different boundary conditions and computational domain dimensions on modeling and simulations of periodic metamaterial arrays in microwave frequencies." *International Journal of RF and Microwave Computer-Aided Engineering* 23, no. 4 (2013): 459-465.

[45] Velez, Paris, Katia Grenier, Javier Mata-Contreras, David Dubuc, and Ferran Martín. "Highly-sensitive microwave sensors based on open complementary split

ring resonators (OCSRRs) for dielectric characterization and solute concentration measurement in liquids." *IEEE Access* 6 (2018): 48324-48338.

[46] Velez, Paris, Jonathan Munoz-Enano, Katia Grenier, Javier Mata-Contreras, David Dubuc, and Ferran Martín. "Split ring resonator-based microwave fluidic sensors for electrolyte concentration measurements." *IEEE Sensors Journal* 19, no. 7 (2018): 2562-2569.

[47] Vrba, Jan, and David Vrba. "A Microwave Metamaterial Inspired Sensor for Non-Invasive Blood Glucose Monitoring." *Radioengineering* 24, no. 4 (2015).

[48] Bakır, Mehmet, Muharrem Karaaslan, Emin Unal, Faruk Karadag, Fatih Özkan Alkurt, Olcay Altıntaş, Sekip Dalgac, and Cumali Sabah. "Microfluidic and fuel adulteration sensing by using chiral metamaterial sensor." *Journal of The Electrochemical Society* 165, no. 11 (2018): B475.

Chapter 5

Polarization Independent Multiple-Band Metamaterial Absorber for Terahertz Applications

Komalpreet Kaur* **and Amanpreet Kaur**
Thapar Institute of Engineering and Technology (TIET), Department of Electronics and Communication Engineering, Patiala, India

Abstract

Multiple band absorbers with metamaterial properties are imperative candidates for many applications, but their accomplishment process is pretty complex. Thus the current chapter proposes a compact size multiple band (Six bands) metamaterial absorber (MA) for terahertz applications. The proposed absorber design consists of three resonators i.e., square with slotted plus-shaped geometry that is enclosed by an outer square loop structure. In addition, a plus-shaped resonator is placed in slotted plus-shaped geometry. The placement of the respective resonator on the substrate produces a capacitive coupling. Thus inducing coupling among these structures provides a compact size MA structure with multiband resonance. All the resonators are made up of gold (Au) metal of 0.5 μm. Each resonator is responsible for the overall performance of the MA. The well-optimized dimension of the unit cell structure is $120 \times 120\ \mu m^2$ imprinted on a substrate with a refractive index $1.73 + 0.1i$ (Porous silicon dioxide with 85% porosity) of thickness 16 microns. The proposed MA design provides absorption rate of 99.7%, 99.8%, 83%, 99%, 98.5% and 98.5% at 0.4, 1.049, 1.9, 2.45, 3.07 and 4 THz resonance frequencies respectively at normal incidence of electromagnetic (EM)

* Corresponding Author's Email: preet.komal0094@gmail.com.

In: Metamaterials for Microwave and Terahertz Applications
Editors: B. Appasani, O. Prakash Acharya, A. Vidyakant Jha et al.
ISBN: 979-8-88697-330-3

wave on the structure. The metamaterial properties of the proposed MA have been verified based on effective permittivity and permeability's negative values at the resonant frequencies. The stability of the proposed MA structure is authenticated through different angles of incident and with different angles of polarization. Furthermore, the multiple frequency range, high absorption, simple design, and compact size make it appropriate for different eminence applications such as stealth technology, surveillance applications, terahertz radar imaging, terahertz antenna systems, etc.

Keywords: metamaterial absorber (MA), permittivity, permeability, polarization-insensitive, stealth applications

1. Introduction

At present, the ordinary absorbers have been replaced by artificial absorbers that efficiently absorb the incoming electromagnetic wave that is prominently identified as metamaterial absorbers (MA) [1-2]. Metamaterial absorber structures are well known as artificial materials because of their unusual characteristics that do not naturally exist in the environment. As per the literature survey, the MA has been broadly examined for various applications that have started from the sub GHz to the optical frequency ranges [3]. These special type absorbers are famous for their selective properties like controllability, adaptability, perfect absorption with negative permittivity and permeability etc., [1-3]. They continually grab the attention of researchers because of their nearly perfect absorption of the incoming Electromagnetic waves, providing adaptable operating performance with ultrathin dielectric substrate at the required frequency range [1, 4-5]. Considering their these exceptional properties, metamaterial absorbers may be the best candidate for the application such as selective thermal emitter, solar energy harvesting [6], surface-enhanced Raman scattering [7], refractive index sensing [8], improving the gain of millimeter antenna, stealth technology, terahertz radar imaging and also in surveillance applications, etc., [3-5]. MA offers more benefits than conventional absorbers, such as miniaturization, wider adaptability, increased effectiveness, etc. The basic idea behind the MA is to design the unit cell structure with novel electric and magnetic properties for incoming electromagnetic waves. According to the literature survey, ultrathin multiband metamaterial absorbers have many impending applications such as these MA are widely used in industry and defense technology [9-11]. The

MAs' unit cell structure typically consists of three layers: the top metallic resonating layer and metallic base layer are separated by the insulating layer (substrate layer). All three layers have important roles in realizing the nearly ideal absorption for the incoming incident wave. Based upon the literature study, the first reported metamaterial is a polarization independent absorber with a narrow-band frequency response. The first MA was hypothetically explored in 2006 based on the array of split-ring resonators (SRRs) [12]. When the incoming EM wave impinges parallel to the SRR structure, its magnetic field is perpendicular to the SRR array. The array is arranged at the resistive sheet to maintain the free space impedance matching. Thus both the reflection coefficient (S11) and the transmission coefficient (S21) of the response are beneath −20 dB, and the locality of the resonance peak is found mathematically. Thus perfect absorption curve had been achieved for this structure because of its strong resonance. Though the bandwidth range of the absorption curve was a little low, such a design motivates further research in this field. In 2008, Landy et al., [1] designed a MA structure from an electric ring as a resonator that was imprinted on an FR-4 substrate. In addition, a cut wire element is imprinted on the flip side of the substrate. This was the first MA with a planar and profoundly sub-wavelength structure. The absorption rate was nearly 99% with the simulation and 88% for measured results at 11.5 GHz. Initially, MAs were polarization-sensitive because of the unsymmetrical unit cell [13]. Subsequently, polarization-insensitive metamaterial absorbers were developed with symmetrical shapes, such as circular resonators [14], ring-shaped, snowflake [15-16] cells, etc. In 2009, Zhu et al., proposed dendritic unit cell-based MA imprinted on the front side of the FR- 4 and another side of the substrate fully ground [17]. The design offered 95% absorption for different polarization angles.

Besides, multi resonant metamaterial absorbers were later established by combining the effect of multi resonant metallic elements using coplanar strategy. In addition, the structure with multiple close resonance absorption peaks can be realized to achieve broadband absorption [18]. In addition, the broadband absorption from the unit cell structure can be achievable using stack form. Furthermore, the absorptivity and the resonance frequencies of metamaterial absorbers can be tuned via external biases by simply adding an active medium such as diodes.

This chapter presents a multiband MA for multiple applications. The fundamental cell of the multiband MA is framed by a cross-shaped geometry further enclosed by a metallic (Au) square loop put on a dielectric sheet. Multiple narrow band absorptions curves with absorption 99.7%, 99.8%, 83%,

99%, 98.5% and 98.5% has been achieved at frequency 0.4, 1.049, 1.9, 2.45, 3.07 and 4 THz respectively. The given MA can find many applications in THz sensing.

2. Geometry and Design Procedure

The approach for designing a MA for a specific frequency band characteristic starts with its unit cell design; this unit cell is then replicated as a periodic structure to make a final prototype sheet with desired band characteristics. Figure 1(a-b) depicts the perspective and top view of the proposed unit cell of MA to achieve the multiple resonances for terahertz application. The sandwich structure of resonator (Au), insulator (substrate) and the bottom ground plane of Au is the study object of this article. The resonator of the design consists of three sub resonators responsible for the overall performance of the MA. The proposed MA unit cell is imprinted on a dielectric substrate with a dielectric constant of 1.73 + 0.1i and a height (h) of 16 μm. The unit cell design consists of an outer thin square loop of gold (of dimensions 116 × 116 × 0.5μm3) with a thickness of 6 μm imprinted on the top of the substrate. In addition, a square-shaped structure with slotted plus is enclosed by the square loop, as shown in Figure 1(b). Each metallic geometry of the resonator is responsible for the overall multiband metamaterial absorber's performance. A plus-shaped structure (with a horizontal and vertical leg of dimension 60×30 μm2) is enclosed in the slotted plus geometry that further sharpens the response of the proposed MA in terms of absorption and stability at incidence angles, as shown in Figure 2(d). The design procedure of the MA is carried out using a 3D EM tool and simulated using CST Microwave Studio software. Floquet port boundary conditions are applied. However, the absorption rate of the proposed design is calculated by equation (1) where T(ω) represents the transmission parameters of the proposed MA and R(ω) represents the reflection parameters of the proposed MA.

When an electromagnetic wave of a wide frequency band impinges on the proposed MA surface, it offers the required absorption for multiple bands in the THz frequency range. Hence the proposed MA is suitable for different quality applications such as stealth technology, surveillance applications, terahertz radar imaging, antenna systems, and other EM devices.

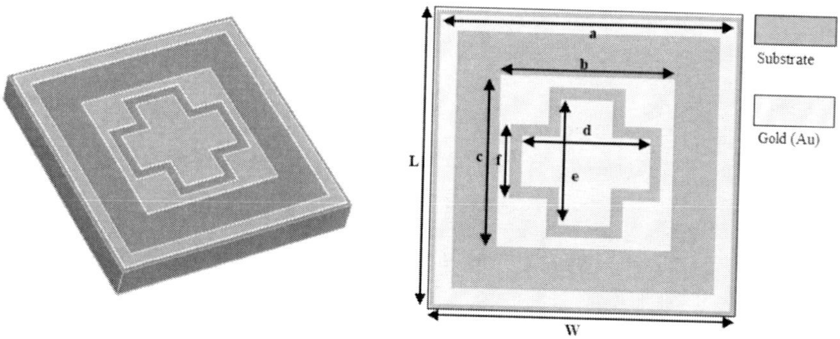

(a) (b)

Figure 1. (a) Perspective view and (b) top view of the proposed metamaterial absorber.

To deeply examine the operation of the proposed MA design, the designing process is explained in three steps (Figure 2(a)), and their corresponding absorptions curves are shown in Figure 2(b-d). Figure 2 (a) depicts the different steps for the evolution of the proposed MA. In the first step (Figure 2(a)), only a square loop is imprinted on the front side of the substrate, and another side of the ground is fully gold plated with a thickness of 0.5 µm. Figure 2(b) depicts the absorption curve for step 1(Figure 2(a)). It is seen that the proposed MA unit cell structure offers one major absorption curve at frequency 0.4 THz with an absorption rate of 99%. It is also observed that the imprinted metallic square loop also offers sub-absorption curves with absorption rates of 99%, 30%, 60%, 40%, and 55% at frequencies 1.9, 2.4, 3.06, and 4.33 THz, respectively. Moreover, these absorption rates are further improved with the involvement of sub resonators, as shown in Figure 2(b-d). In step 2 (Figure 2(a)) slotted cross-shaped geometry is enclosed by a square loop, and the combined effect of both the resonator on absorption curves is shown in Figure 2(c). This arrangement of the two different resonators shifts the resonance peak from 1.9, 3.06, and 4.33 THz to 1.05, 2.85, and 4.01 THz, respectively, as shown in Figure 2(c). Interestingly, the absorptions for the shifted resonance peaks have improved up to 98%, 95%, 90%, and 97% at 1.05, 2.42, 2.85, and 4.01 THz, respectively. Finally, a plus-shaped structure is enclosed by the geometry of step 2 resonator for obtaining multiband absorption curves, as shown in step 3 (Figure 2(a)). It is seen from Figure 2(d) that this resultant arrangement of three sub resonators over the top side of the substrate offers multiband (six resonance peaks) absorption curves.

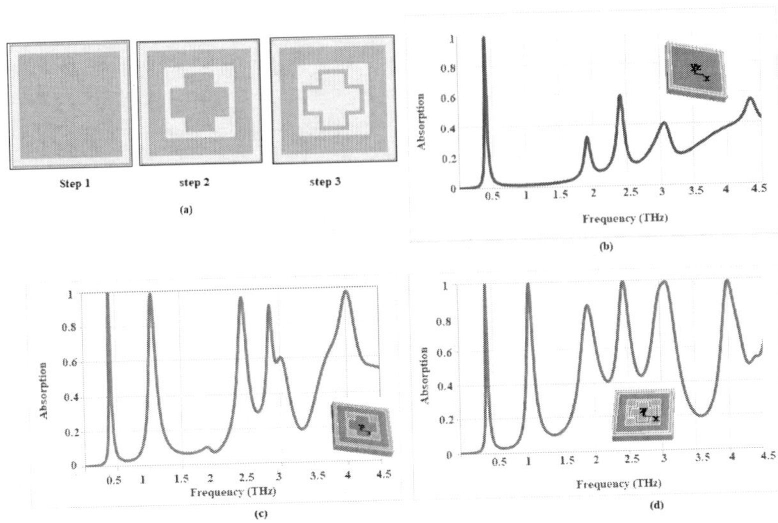

Figure 2. (a) Evolution steps for proposed MA, (b-d) absorption curves for (b) step 1, (c) step 2 and (d) step 3 respectively.

Additionally, the combined structure adds one more resonance absorption at 1.9 THz, with absorption of approximately 83%. It is observed that it not only introduces the coupling between both geometries but also provides strong capacitance between them. It is seen that the placement of the crossed-shaped geometry inside the resultant structure provides the strong mutual coupling and provides six frequency bands at frequencies 0.4, 1.049, 1.9, 2.45, 3.07, and 4 THz with absorption 99%, 99%, 83%, 99%, 98.5%, and 98.5% respectively. Thus final structure resonates at frequencies 0.4, 1.049, 1.9, 2.45, 3.07 and 4 and their corresponding wavelengths are 0.75, 0.285, 0.157, 0.122, 0.097 and 0.075mm as shown in Figure 3(a) and 3(b) respectively. Figure 3(a) also shows the combined results of absorption rate and reflection parameters (S_{11}) of the proposed absorber.

Moreover, the ingestion of the unit cell MA is characterized as [18]:

$$A(\omega) = 1 - T(\omega) - R(\omega) \tag{1}$$

where, $A(\omega)$ represents the absorption of the proposed design, $R(\omega)$ represents the reflection, and $T(\omega)$ indicates the transmittance of the MA. In addition, the transmission coefficients are diminished once the dielectric properties such as effective permittivity $\varepsilon(\omega)$ and permeability $\mu(\omega)$ are nearly the same. Thus, the normalized impedance Z is effortlessly well matched with the free space impedance for these frequencies. Subsequently, the absorption of the proposed MA is maximized. The standardized impedance is determined using equation (2) [19].

$$z = \sqrt{\frac{(1+S_{11})^2 - S_{21}^2}{(1-S_{11})^2 - S_{21}^2}} \tag{2}$$

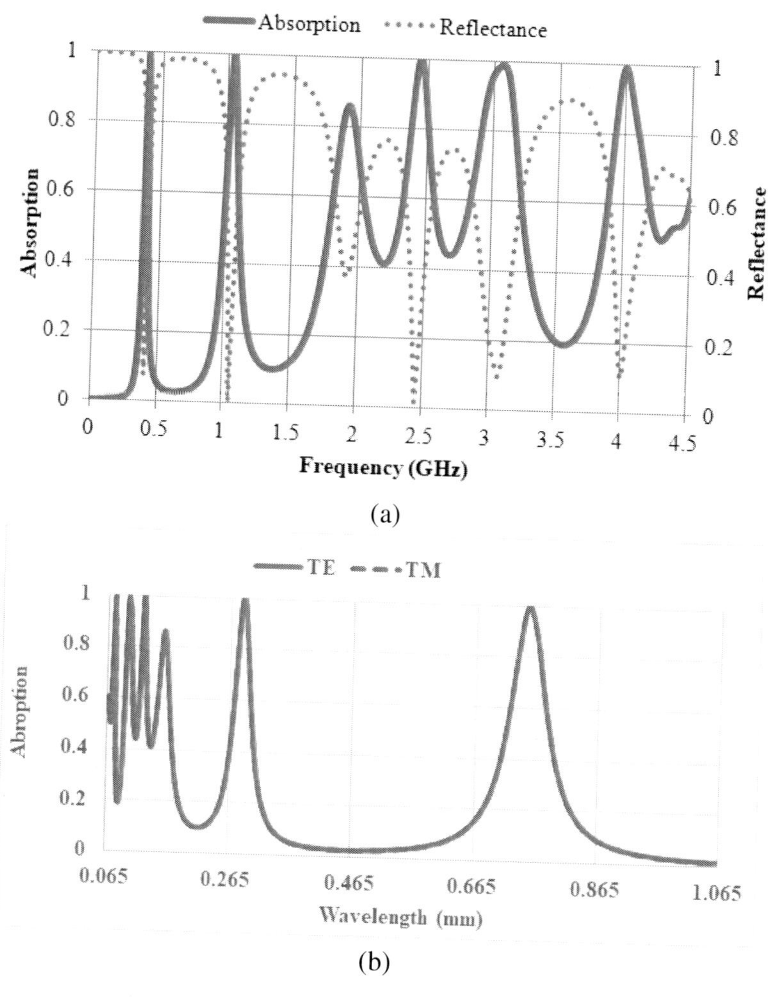

Figure 3. (a) Absorption and reflection parameters for Proposed MA as a function of frequency, (b) Absorption rate as a function of wavelength

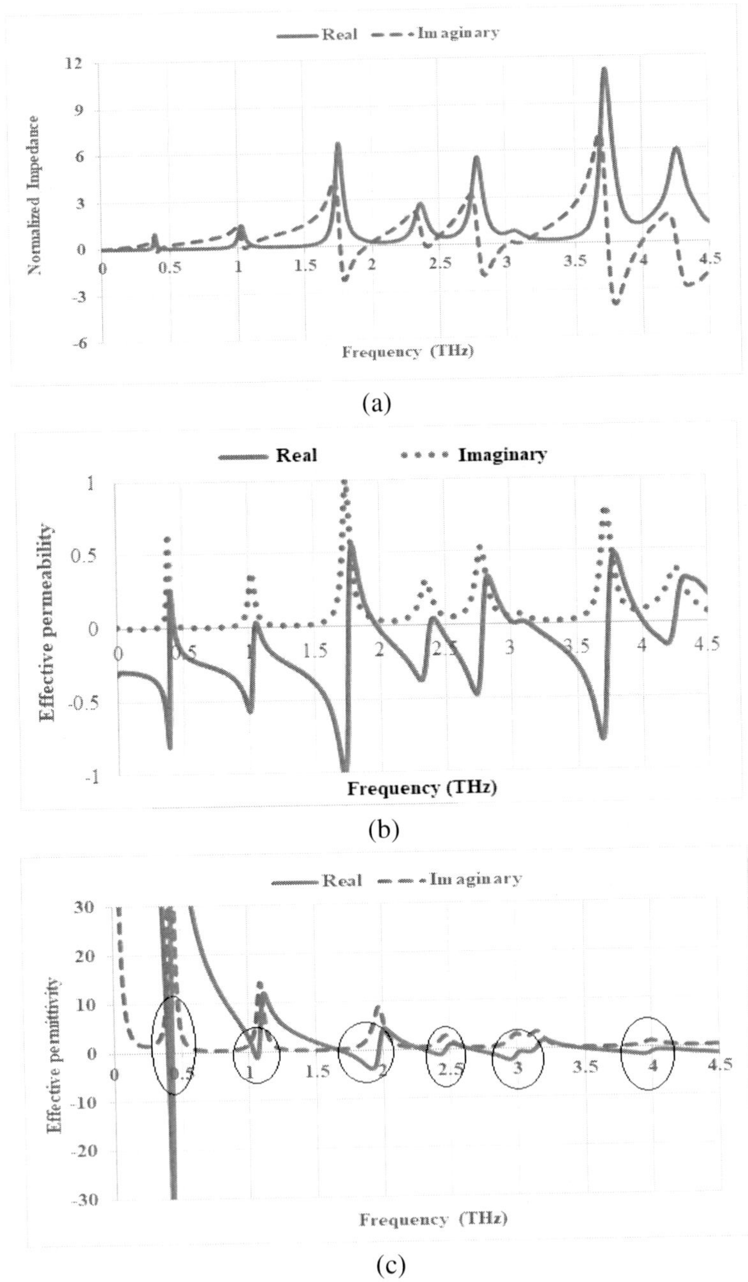

Figure 4. (a) Normalized impedance, (b) effective permeability and (c) effective permittivity of the proposed multiband MA.

Figure 4 (a) demonstrates the real (solid blue line) and imaginary (dotted brown line) portion of normalized input impedance. It is seen that calculated real and imaginary values of the input impedance Z are nearly equivalent to zero and one at the resonance frequencies. Moreover, the metamaterial attributes, the effective permittivity (ε_{eff}) [20], and permeability (μ_{eff}) [20] of the proposed MA have been found with the help of equations (3) and (4), respectively.

$$\xi_{eff} = 1 + \frac{2i \times (1 - S_{11} - S_{21})}{k_0 t \times (1 + S_{11} + S_{21})} \tag{3}$$

and

$$\mu_{eff} = 1 + \frac{2i \times (1 + S_{11} - S_{21})}{k_0 t \times (1 - S_{11} + S_{21})} \tag{4}$$

where ε_{eff} and μ_{eff} are effective permittivity and effective permeability of the MA. k_0 is the wave number in the free space, and t is the dielectric substrate depth used as a sandwich between metallic layers. Figure 4(b-c) depicts the calculated real and imaginary parts of the complex effective permittivity and permeability, respectively. It is seen that these characteristics offer some negative values at the resonance peaks. Thus these negative values prove the metamaterial properties of the proposed absorber structure.

3. Results and Discussion

The parameters of the proposed MA have been optimized to improve the performance of the MA.

3.1. Parametric Study

The parametric investigation is done by computing the reflection coefficient and the transmission coefficient at the normal incidence of the incoming plane wave. By altering the dimensions of the unit cell element, the performance of the MA in absorption rate varies at various resonant frequencies. At first, parameter b is varied by keeping the rest of the parameters fixed, and its corresponding absorption curves are shown in Figure 5(a). It is observed that with the increases in the length of the structure, resonance peaks shift towards the higher frequency side and reduce the absorption rate. It is also interesting to note that beyond the length of 70 μm, there is a sudden reduction in absorption at the higher frequency. Thus the best absorption curves are found at a length of 68 μm, and therefore, it is selected.

Furthermore, the effect of length d of plus structure on the multiband absorption curves is shown in Figure 5(b). Within the increase of the horizontal and vertical length of the plus-shaped structure, absorption curves shift towards the high-frequency side. In addition, at the length of 56 μm, there is a reduction in the absorption at 3.07 and 4.0 THz, respectively. Moreover, with the increases in length up to 58 μm, there is an improvement in the absorption rate near 3.07 THz, but still, there is low absorption at 4.0 THz, as shown in Figure 5(b). Similarly, at the length of 62 μm, the absorption rate has been improved for each resonance, but the absorption curve is not smooth. Thus, the best absorption curves are observed at 60 μm, which is the optimized parameter of the proposed MA.

The effect of the thickness on the performance of the MA is shown in Figure 5(c). It is seen that with the increase in the thickness t of the outer loop of the MA structure, there is an increase in the inductance. Thus the resonance frequencies shift towards the lower frequency side. It is also observed that the absorption rate at the resonance peaks is reduced by enhancing the thickness of the outer square loop structure. In addition, for t between 8 μm and 10 μm, the absorption rate of the proposed MA is reduced below 85% at 3.02 THz and 4.2 THz frequencies. Finally, the desired absorption rate for each resonance peak is observed at the thickness t = 6 μm; therefore, it is selected.

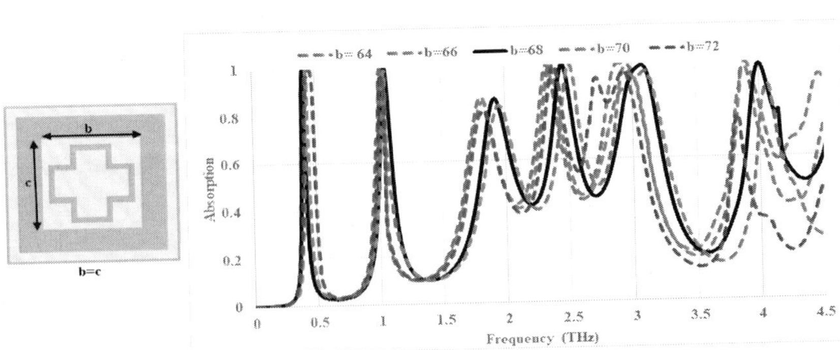

Figure 5. (a) Effect of length b of slotted structure on the absorption.

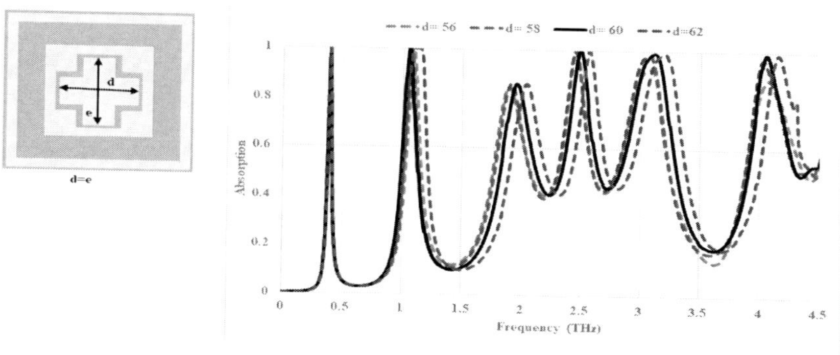

Figure 5. (b) Effect of length at *d* of plus structure on the absorption.

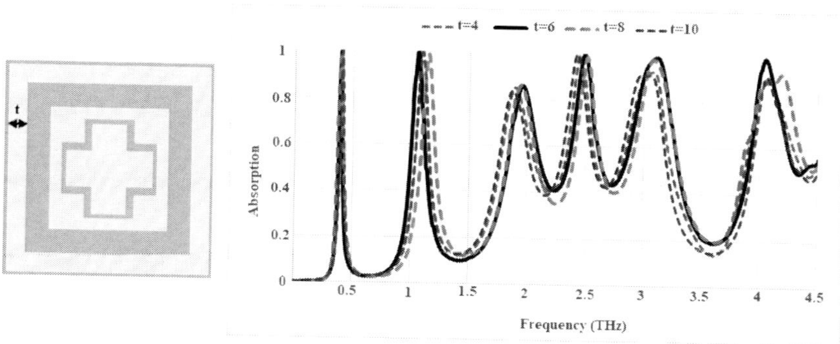

Figure 5. (c) Effect of thickness *t* of outer square loop structure on the absorption.

3.2. Absorption at Different Incidence and Polarization Angles

The simulated outcomes for the variety in absorption rate with altering the angle of incidence (θ) for both TE mode and at a different angle of polarization for both TE and TM mode have been shown in Figure 6(a-c), respectively. For the TE mode of polarization, the direction of electric field vector E is steady alongside the x-axis, while the direction of both magnetic field vector H and wave vector varies by different incidence angle (θ). In addition, it is observed that there is an insignificant shift in the resonance peak of the proposed MA at a lower frequency, with the variation in incidence angle. It is also seen that the varying angle does not affect the absorption rate of the resonance peaks, as shown in Figure 6(a). It is interesting to notice that varying angle θ affects the absorption rate of higher resonance frequencies more than lower frequencies.

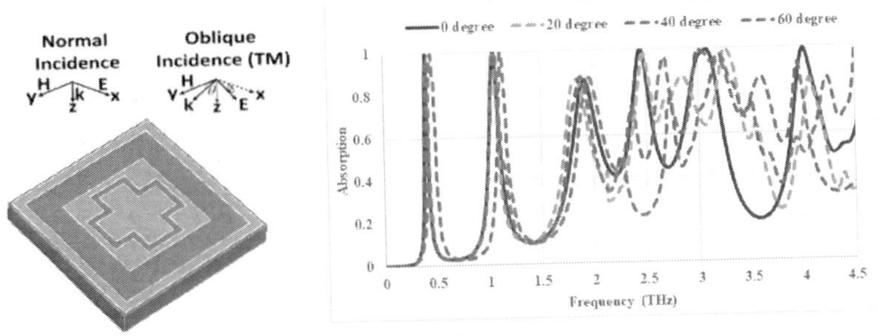

(a)

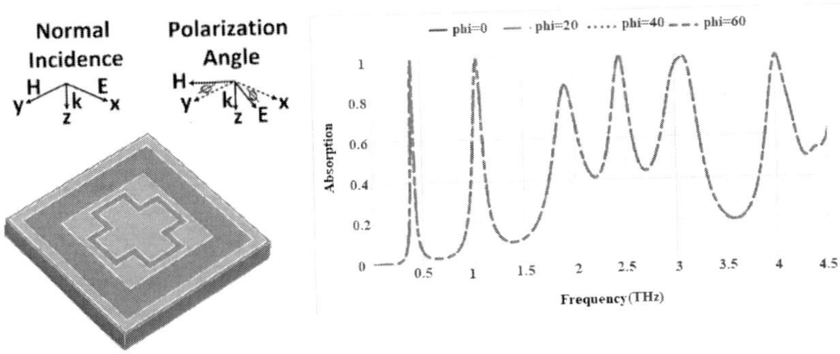

(b)

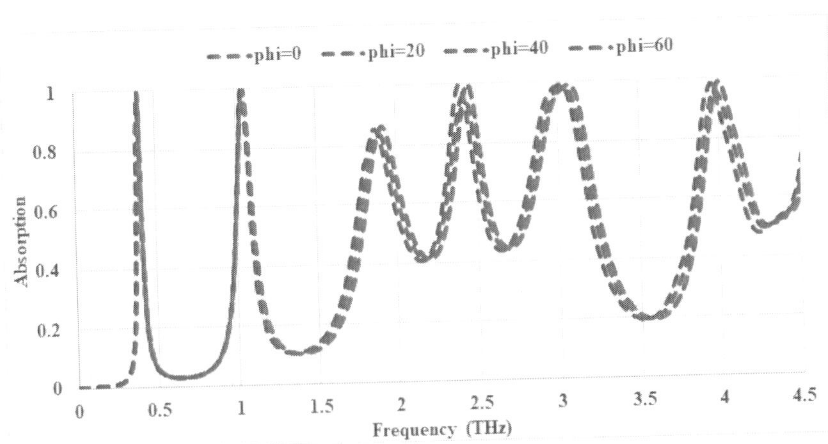

(c)

Figure 6. (a) Absorption curves of the proposed MA at different incidence angles (θ) for TE polarized wave, (b) Absorption curves of the proposed MA at different polarization angles (ϕ) for TE mode (c) Absorption curves of the proposed MA at different polarization angles (ϕ) for TM mode.

Also, the absorption rate remains the same up to 60° for frequencies 0.4, 1.049, and 1.9 THz, while some unwanted resonance peaks appear for those at 2.45, 3.07, and 4 THz frequencies, as shown in Figure 6(a). In addition to the study of angle of incidence, the study of polarization angle over its performance is also important. Thus the proposed MA has been also investigated for both TE and TM modes for different angles of polarization (ϕ). For the TE mode, the direction of the propagating wave remains alongside the z-axis while the electric field vector (E) and magnetic field vector (H) made an angle phi (ϕ) along axis X and Y respectively. The absorption rate remains unaffected by the polarization angle as shown in Figure 6(b-c). As the unit cell element of the proposed MA is symmetrical, there is an insignificant difference in the absorption curves at varying polarization angles from $\phi = 0°$ to 60°. Similarly, absorption curves for the TM mode of polarization for different polarization angles have been shown in Figure 6(c). There is a small variation in the resonance peaks of the proposed MA at $\phi = 60°$. However, the absorption rate remains almost the same. Thus, the proposed metamaterial absorber offers polarization independence for both TE and TM modes.

Conclusion

The present chapter introduces a compact six-band ultrathin metamaterial absorber. The proposed MA has gold-plated resonators imprinted on the dielectric substrate of refractive index 1.73 + 0.1i, with dimensions 120 ×120 μm^2 and a thickness of 16 µm. Moreover, the flip side of the substrate has a 0.5 µm thick gold layer. The proposed MA is resonating at 0.4, 1.049, 1.9, 2.45, 3.07 and 4 THz with absorption rate of 99.7%, 99.8%, 83%, 99%, 98.5% and 98.5% respectively. The properties of the proposed absorber as metamaterial have been verified with negative permittivity and permeability values at resonance peaks. The parametric sensitivity of the designed MA has been perceived by altering specific parameters. Furthermore, the effect of polarization and incident angles on the performance of MA has been studied, and it is observed that the design is insensitive to variations in these angles. The designed MA can be utilized in stealth technology, surveillance applications, terahertz radar imaging, antenna systems, and other EM devices.

References

[1] Landy, N. I., S. Sajuyigbe, Jack J. Mock, David R. Smith, and Willie J. Padilla. "Perfect metamaterial absorber." *Physical review letters* 100, no. 20 (2008): 207402.

[2] Soukoulis, Costas M., and Martin Wegener. "Past achievements and future challenges in the development of three-dimensional photonic metamaterials." *Nature photonics* 5, no. 9 (2011): 523-530.

[3] Cao, Z. X., F. G. Yuan, and L. H. Li. "A super-compact metamaterial absorber cell in L-band." *Journal of Applied Physics* 115, no. 18 (2014): 184904.

[4] Watts, Claire M., Xianliang Liu, and Willie J. Padilla. "Metamaterial electromagnetic wave absorbers." *Advanced materials* 24, no. 23 (2012): OP98-OP120.

[5] Cui, Yanxia, Yingran He, Yi Jin, Fei Ding, Liu Yang, Yuqian Ye, Shoumin Zhong, Yinyue Lin, and Sailing He. "Plasmonic and metamaterial structures as electromagnetic absorbers." *Laser & Photonics Reviews* 8, no. 4 (2014): 495-520.

[6] Wang, Hao, Vijay Prasad Sivan, Arnan Mitchell, Gary Rosengarten, Patrick Phelan, and Liping Wang. "Highly efficient selective metamaterial absorber for high-temperature solar thermal energy harvesting." *Solar Energy Materials and Solar Cells* 137 (2015): 235-242.

[7] Culhaoglu, Ali E., Andrey V. Osipov, and Peter Russer. "Mono-and bistatic scattering reduction by a metamaterial low reflection coating." *IEEE transactions on antennas and propagation* 61, no. 1 (2012): 462-466.

[8] Guddala, Sriram, Raghwendra Kumar, and S. Anantha Ramakrishna. "Thermally induced nonlinear optical absorption in metamaterial perfect absorbers." *Applied Physics Letters* 106, no. 11 (2015): 111901.

[9] Mao, Zhiwen, Shaobin Liu, Borui Bian, Beiyin Wang, Ben Ma, Lin Chen, and Junyi Xu. "Multi-band polarization-insensitive metamaterial absorber based on Chinese ancient coin-shaped structures." *Journal of Applied Physics* 115, no. 20 (2014): 204505.

[10] Chaurasiya, Devkinandan, Saptarshi Ghosh, Somak Bhattacharyya, Anamiya Bhattacharya, and Kumar Vaibhav Srivastava. "Compact multi-band polarisation-insensitive metamaterial absorber." *IET Microwaves, Antennas & Propagation* 10, no. 1 (2016): 94-101.

[11] Ozden, Kadir, Okan Mert Yucedag, and Hasan Kocer. "Metamaterial based broadband RF absorber at X-band." *AEU-International Journal of Electronics and Communications* 70, no. 8 (2016): 1062-1070.

[12] Bilotti, Filiberto, Luca Nucci, and Lucio Vegni. "An SRR based microwave absorber." *Microwave and Optical Technology Letters* 48, no. 11 (2006): 2171-2175.

[13] Diem, Marcus, Thomas Koschny, and Costas M. Soukoulis. "Wide-angle perfect absorber/thermal emitter in the terahertz regime." *Physical Review B* 79, no. 3 (2009): 033101.

[14] Wei-Ren, Zhu, Zhao Xiao-Peng, Bao Shi, and Zhang Yan-Ping. "Highly symmetric planar metamaterial absorbers based on annular and circular patches." *Chinese Physics Letters* 27, no. 1 (2010): 014204.

[15] Huang, Y-J., G-J. Wen, J. Li, W-R. Zhu, Ping Wang, and Y-H. Sun. "Wide-angle and polarization-independent metamaterial absorber based on snowflake-shaped configuration." *Journal of Electromagnetic Waves and Applications* 27, no. 5 (2013): 552-559.

[16] Huang, Yongjun, Yiran Tian, Guangjun Wen, and Weiren Zhu. "Experimental study of absorption band controllable planar metamaterial absorber using asymmetrical snowflake-shaped configuration." *Journal of Optics* 15, no. 5 (2013): 055104.

[17] Zhu, Weiren, and Xiaopeng Zhao. "Metamaterial absorber with dendritic cells at infrared frequencies." *JOSA B* 26, no. 12 (2009): 2382-2385.

[18] Wanghuang, Tenglong, Weijian Chen, Yongjun Huang, and Guangjun Wen. "Analysis of metamaterial absorber in normal and oblique incidence by using interference theory." *Aip Advances* 3, no. 10 (2013): 102118.

[19] Bhattacharyya, Somak, Saptarshi Ghosh, and Kumar Vaibhav Srivastava. "Triple band polarization-independent metamaterial absorber with bandwidth enhancement at X-band." *Journal of Applied Physics* 114, no. 9 (2013): 094514.

[20] Sarkhel, Abhishek, and Sekhar Ranjan Bhadra Chaudhuri. "Compact quad-band polarization-insensitive ultrathin metamaterial absorber with wide angle stability." *IEEE Antennas and Wireless Propagation Letters* 16 (2017): 3240-3244.

Chapter 6

Terahertz Metamaterial Filters: Design, Fabrication, and Applications

L. Harlan[1,*], Sachin Kumar[1,†] and Shyamal Mondal[2,‡]

[1]Department of Electronics and Communication Engineering,
SRM Institute of Science and Technology, Kattankulathur, India
[2]Department of Applied Physics, Defence Institute of Advanced Technology, Pune, India

Abstract

Research in the terahertz (THz) frequency range is thriving with advancements in semiconductor technology and nano-fabrication engineering. Metamaterials (MMs), due to their unique properties, allow effective control of radiations in the THz and are becoming popular. The resonant frequency can be determined by examining the artificially engineered MMs and their structural cell geometry. Electromagnetic properties such as permeability and permittivity can be altered by changing their structural cell geometry, which leads to designing filters in the THz regime. The THz spectrum (0.1–10 THz), which is non-ionizing, has some interesting properties such as penetration through a wide variety of materials and selectively absorbed by water and organic substances. Moreover, many materials reveal their distinctive spectral lines at this range, which provide unique information about their structure. Many systems for imaging, sensing, and wireless communications have been developed by inheriting the properties of THz radiation. Along with other components in the THz regime, it is essential to design a filter in the THz range that can select the spectrum to be sent

[*] Corresponding Author's Email: hl0359@srmist.edu.in.
[†] Corresponding Author's Email: sachinkr@srmist.edu.in.
[‡] Corresponding Author's Email: shyamal.kgec@gmail.com.

In: Metamaterials for Microwave and Terahertz Applications
Editors: B. Appasani, O. Prakash Acharya, A. Vidyakant Jha et al.
ISBN: 979-8-88697-330-3

and make the communication system more flexible. In this chapter, the theory of metamaterials that can work in the THz regime has been discussed. The designs and operating principles of different types of MM-based THz filters have been described. The fabrication for the THz filters has also been described. The results of published works have been discussed and compared, along with their potential applications.

1. Introduction to Terahertz Metamaterials

The metamaterial-based terahertz filters (THz MMFs) provide enough flexibility for researchers to exploit the maximum bandwidth in the THz regime. The advancement of micro-and nano-fabrication technologies opens up lucrative opportunities to use THz MMFs in various applications, including biomedical, space, security, defense, etc. Since the invention of metamaterials by J. Pendry et al. [1], various studies have been conducted on these artificial materials due to their unique properties to shape the electromagnetic spectrum. In 2004, J. Yen et al. showed that a planar structure at THz frequencies could be used to manipulate the magnetic response. They reported a non-magnetic conductive resonant element, an artificial magnetic or hybrid structure capable of changing the properties of THz waves [2]. Filters in the THz regime, like in other electromagnetic domains, play an important role. In order to understand the working principle of THz metamaterial filters, it is better to know the metamaterial classification and the various types of metamaterial structures based on it. In 2006, N. Engheta and R. W. Ziolkowski [3] classified metamaterials (MMs) into two major structures based on effective permittivity (ε) and permeability (μ); first, the double negative (DNG) structures, where both ε and μ are negative, and the single negative (SNG) structures, where any one of ε and μ is negative. Second, the photonic band gap (PBG) structures, where the periodic refractive index profile realizes the metamaterial properties. These properties can be realized using several configurations such as split-ring resonator (SRR), frequency selective surface (FSS), artificial magnetic conductor (AMC), high impedance surface (HIS), etc. The details of typical metamaterial structures are shown below.

1.1. Split Ring Resonator

Natural magnetism is difficult to find in materials that can be used in wideband frequencies (beyond a few Gigahertz). J. Pendry proposed a solution to this problem in 1999 with the invention of an artificial non-magnetic conductor with a magnetic response known as the SRR. Over the past few decades, SRR has shown promising results in creating an effective negative permeability in the THz frequency range.

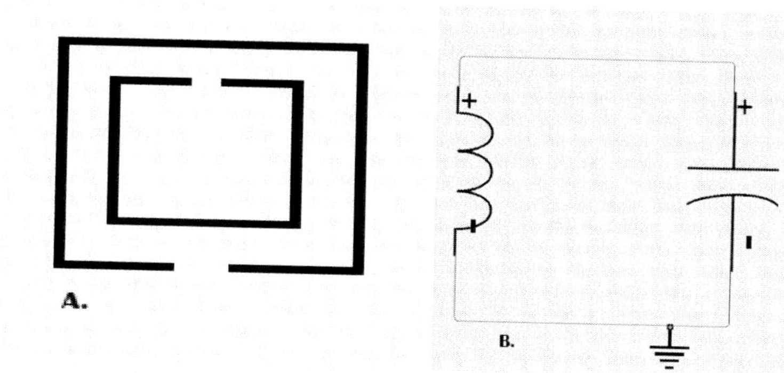

Figure 1. (a) SRR structure and (b) SRR Equivalent circuit. Reproduced from [4].

The SRR is made up of metallic structures in different geometrical shapes printed on the dielectric substrate. The metallic structures, usually in the form of rings, are made of non-magnetic materials such as copper, gold, silver, etc., and have a small split (gap) between them. The non-magnetic rings can be single, double, or multiple depending on the application. When the SRR is placed in a time-varying magnetic field, a circular electric current is induced in the metallic rings, and a magnetic dipole moment is produced perpendicular to its plane. The gap offers the capacitance, while the loop offers the inductance. The SRR is realized as an LC resonant circuit, as shown in Figure 1.

The resonance frequency of the *LC* resonant circuit is given by

$$F_r = \frac{1}{2\pi\sqrt{LC}} \qquad (1)$$

where *C* is the capacitance and *L* is the inductance.

These man-made materials can offer surprising electromagnetic responses as they are designed at sub-wavelengths, acting as artificial materials. Such thin-film metamaterials can be customized for different applications due to their selective responses to electromagnetic, acoustic, and thermal waves [4].

1.2. Hybrid Structures

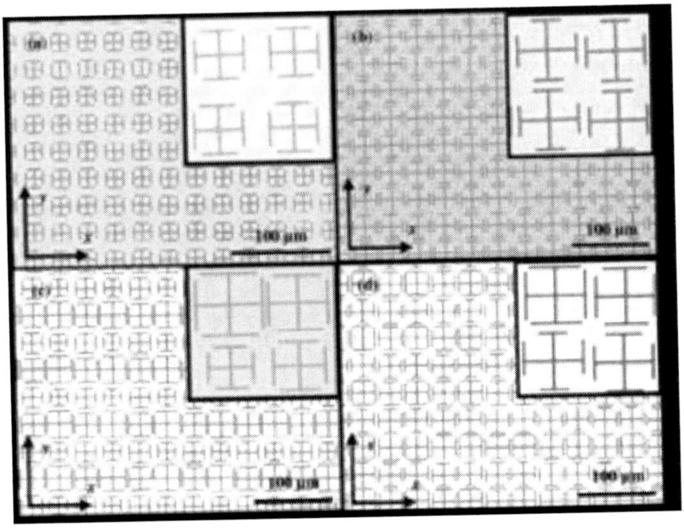

Figure 2. (a) Same core size SRRs, (b) Same antenna gap size SRRs, (c) Hybrid design 1 with constant gap of 3 μm and different core sizes of 24, 28, 32, and 36 μm (d) Hybrid design 2 with constant core size of 32 μm and different gap sizes of 2, 4, 6, and 8 μm. Reproduced from [5].

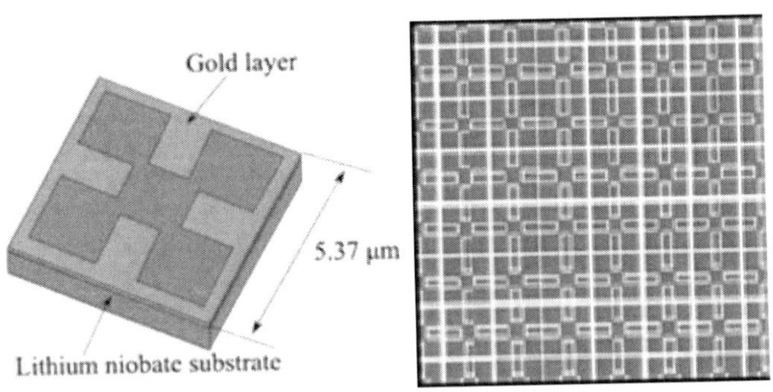

Figure 3. (a) 3D view of a unit cell with a gold outer layer and lithium niobate substrate, and (b) Array of unit cells. Reproduced from [6].

The main motivation in THz communication is obtaining unique THz resonance responses by using metamaterial-based SRR in the sub-THz wavelength. This miniaturization creates numerous opportunities for the realization of novel THz devices for real-world applications, such as filters, absorbers, and modulators. A traditional metamaterial with a single SRR structure produces a single resonance peak at a specific THz frequency. Therefore, multiple resonances at different THz frequencies can be created by fabricating multiple SRR structures of different sizes on the same substrate, and these structures are referred to as hybrid structures. However, the mutual coupling effect between neighboring SRRs is strong and should be considered. Figures 2 and 3 depict the hybrid structure.

1.3. Tunable Terahertz Metamaterial

The application of metamaterials towards THz can be improved by making them more flexible and tunable. Many recent tuning methods have been presented, including electrical tuning, thermal annealing, and laser pumping. However, these methods require a complex mechanism to perform tunable metamaterial. There are various methods for controlling metamaterials: (i) By changing the surrounding medium and (ii) By changing the substrate. The tunability remains limited whether the structure is designed on a flexible or rigid substrate as it depends on the metamaterial's unit cell.

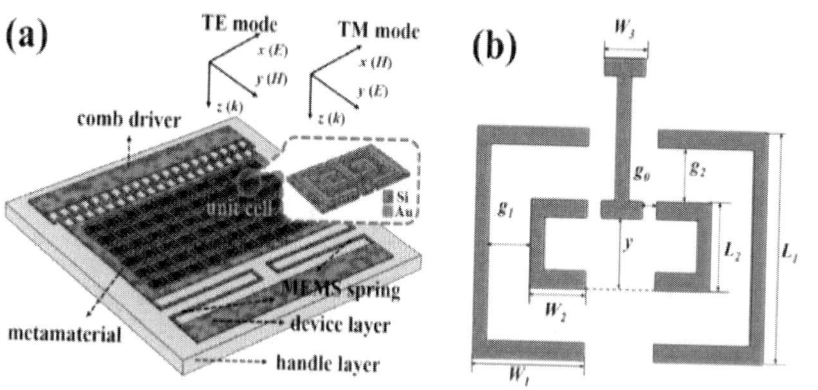

Figure 4. Tunable SRR based on MEMS: (a) 3D view of the MEMS-based SRR, and (b) Dimensions of the unit cell. Reproduced from [12].

Moreover, modifying the MMs unit cell is challenging [7-9]. These limitations can be overcome by introducing micro-electro-mechanical system (MEMS)-based actuators that can change the shape of the metamaterial structures. The MEMS-driven methods can effectively tune the resonant frequency of the metamaterial. MEMS offers a wide range of tunable methods such as electro thermal actuation force, electrostatic actuation force, electro-optical, and magneto-optical forces [9-12]. Despite its advantages, the design of tunable THz metamaterials using MEMS has some serious limitations as most of its actuators require higher driving voltages. However, when compared to the MEMS-based tuning methods, electrostatically based tuning is more promising as it requires less power and reduces thermal noise, providing a greater aspect for imaging and sensing applications. Figure 4 shows an example of a MEMS-based tunable metamaterial structure.

2. Fabrication Process

Fabrication is the most important aspect when designing THz MMF as it requires precision and accuracy [13]. The filter's performance is based on its design precision. Some of the promising fabrication techniques are listed below.

2.1. Deep Reactive Ion Etching

Deep reactive ion etching (DRIE) and mask photolithography (Figure 5a) are the commonly used techniques in the fabrication of silicon-based metamaterial [14, 15]. The fabrication process involves creating a pattern on the photoresist (PR) layer by mask photolithography, followed by DRIE. The steps involved in the fabrication process are described in detail below. A thin oxide layer of silicon is first deposited on top of the silicon wafer using plasma-enhanced chemical vapour, followed by spin coating of PR layer on top of the oxide layer, as shown in Figure 5b. The PR and oxide layer are etched using photolithography and dry etching methods, which act as a mask for DRIE. The oxide mask provides better shielding effects for DRIE because of its excellent oxide selectivity. A metal or glass layer is often used as a substrate at the bottom of the silicon wafer. The parasitic capacitance introduced by the metal layer at the bottom of the substrate is small, but it inhibits charge accumulation. The footing effect can be effectively restrained with the help of a glass wafer. After the DRIE process, the PR layer is removed with isopropanol and acetone, while the oxide and metal layers are leaned with piranha and hydrogen fluoride. The final step is to peel off the silicon-based metamaterial, which can be used for various applications in the THz field. The SEM image shown in Figure 17c has shown the final sample after processing.

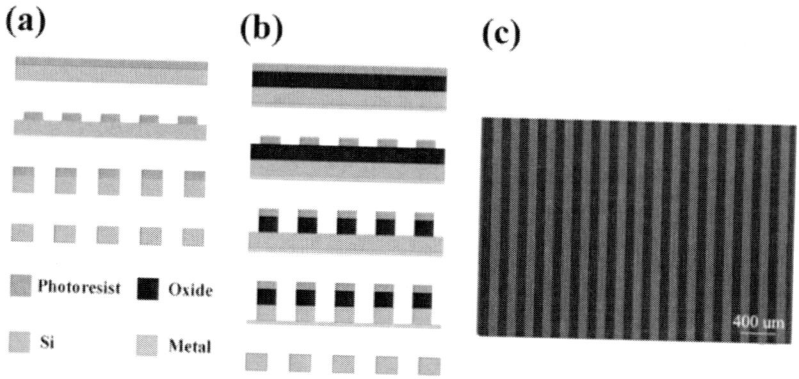

Figure 5. Fabrication process involved in silicon-based MMs: (a) Traditional method, (b) Addition of silicon oxide layer and metal layer on both sides of the silicon wafer, and (c) Sample after processing. Reproduced from [16].

2.2. Laser Micromachining Methods

Laser micromachining is a high tech etching technique that uses both mechanical and optical technologies. It is done to describe the surface shape of quartz or silicon materials for various micro and nanoscale structures. Laser micromachining is extensively used in fabricating materials across a wide range of frequencies (optics to microwave) as it involves less complexity and high precision. Furthermore, it can be used in lithographic processes [17]. Laser micromachining is often used to obtain the desired masks for DRIE processes.

2.3. Self-Assembly Method

The existing fabrication techniques face many cost, precision, and scalability problems. Thus, the micro template assisted self-assembly (MTAS) method has been proposed for large-scale preparation of high-precision flexible structures. The procedures involved in this method are shown in Figure 6 [18]. A thin stainless steel component with varied shapes is first etched using laser micromachining, that is, with a solid-state laser. The micro machined structures serve as a template for MTAS method. The microspheres and steel templates are held together by an adhesive layer. A soft slicker that can blow the microspheres is used to move it from left to right. The microspheres are pushed and locked into the holes of the steel templates by the resultant external force and gravity. Finally, a well-ordered microsphere-based MM is obtained after removing the steel template and remaining microspheres. Figure 6. (b) shows the fabricated sample, which has a regular shape. The fabrication process is extended to design low-cost, large-scale processes as it ensures great flexibility and ecologically friendly fabrication schemes.

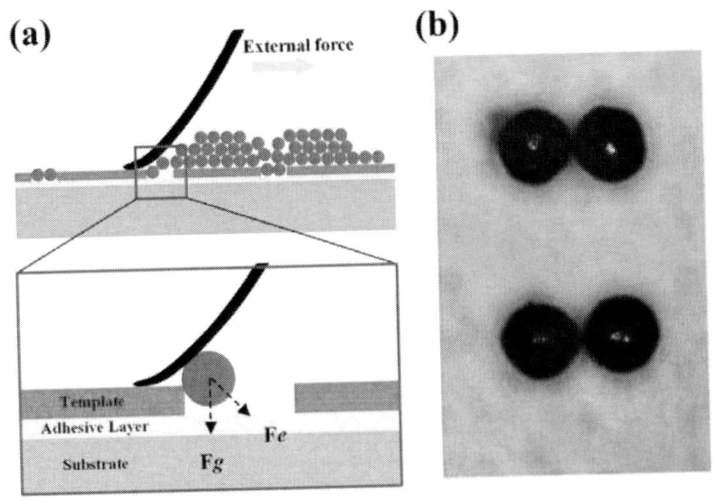

Figure 6. (a) Schematic of MTAS, and (b) Fabricated sample. Reproduced from [18].

2.4. 3D Direct Writing Method

The 3D direct writing technique is the most recent advancement in THz device fabrication. This technique offers low-cost and simple fabrication solutions. The ink for the 3D printer is prepared from $TbFeO_3$ by mixing Tb_4O_7 and pre-sintered Fe_2O_3 powders in a deionized water solution of polyvinyl alcohol (PVA) [19]. A systematic addition of Fe_2O_3, followed by thorough shaking of the mixture for 20 – 30 minutes, is required to improve the ink's quality. The ink is then loaded into the cartridge of the 3D printer. The desired metamaterial structure is then designed using simulation software and 3D printer. A layer-by-layer fabrication procedure for developing a 3D metamaterial design with the help of computer programs is shown in Figure 7. The suggested metamaterial is created using a traditional solid-state sintering process.

Figure 7. Illustration of 3D direct writing technique. Reproduced from [19].

3. Terahertz Metamaterial Filter

Filters are extensively used in various applications such as communication, image processing, sensing, and spectroscopy. The concept of THz metamaterial-based filter (bandpass) was introduced by Lu et al. in 2010 [20]. Since then, various research works have been carried, and different types of THz metamaterial filters have been realized. A few of them are explained below.

3.1. Fabry–Perot Terahertz Filter

Fabry–Perot (FP) optical filters, named after Charles Fabry and Alfred Perot, are well-known for their applications in optical range [21]. A FP filter is a multilayer cavity structure created by repeatedly repeating two different sub-wavelength semiconductor or dielectric materials, resulting in an anisotropic material. The effective refractive index of the medium can be easily modified by varying the fill factor of the metamaterial. Thus, the resonant wavelength can be easily tuned by changing the fill factor as it depends on the effective refractive index of the medium for a wide-ranging wavelength of around 400 nm [22].

FP resonators are used as high-resolution optical filters in a variety of photonic applications, such as spectroscopy. An FP filter functions similarly to a resonant filter, allowing only a particular frequency band to pass through while reflecting all other frequencies. The main components of typical FP resonators are distributed bragg reflectors (DBRs) or a pair of high reflectivity mirrors like metals and a medium filled with a dielectric or semiconductor material [23]. The resonant frequency of the filter can be effectively changed by varying the thickness and dielectric material of the cavity [24-26]. The resonance frequency can also be changed without disturbing the geometry of the filter. The fill factor of the metamaterial is changed in the FP filter shown in Figure 8 to tune its resonant frequency. Also, the filter is made up of two different layers of Si and SiO_2. The filter has application in imaging and spectroscopy.

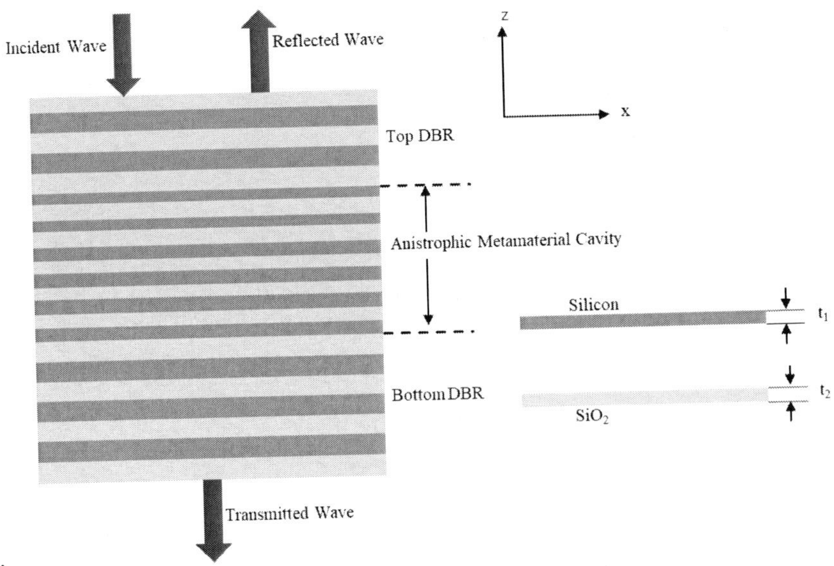

Figure 8. Typical FP filter. Reproduced from [22].

The above structure depicted is composed of alternating layers of Si and SiO_2 that include top and bottom (four pairs each) layers of DBRs with thicknesses of 110 nm and 260 nm, respectively. The cavity is also made up of Si and SiO_2, but is confined with sub-wavelength thicknesses of t_1 and t_2, respectively.

Another interesting tunable FP filter based on all-dielectric metamaterials (ADMs) for the THz regime was presented by T. Ao, et al. Unlike metallic metamaterial reflectors, ADMs are made up of high-permittivity TiO_2 cylinder arrays that exhibit excellent reflectivity over a wide frequency range of 2.49–

3.08 THz. The principle behind the ADMs in the FP filter behaving like reflectors correspond to the Mie resonances (first and second). The ADMs based FP filter has a resonance frequency in the range of 2.78 THz for a cavity with a thickness of $\lambda/18$. Variation of the resonance frequency provides tunability, which can be altered by changing the cavity thickness. The tunable FP filter provides a realistic design and implementation of low-loss THz filters with a thin air cavity.

Figures 9 (a) and (b) compares a typical FP filter made up of two DBRs separated by a half-wavelength air cavity to an ADMs-based FP filter. The device dimensions can be significantly reduced by substituting the DBRs with metallic MMs reflectors. However, the fabrication of metallic MMs reflectors requires high precision, limiting their applications.

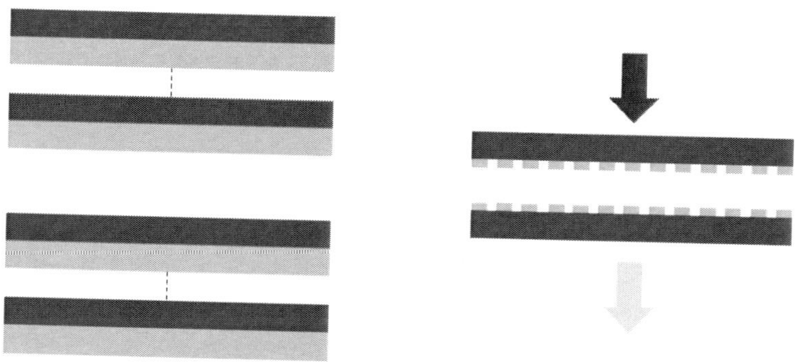

Figure 9. (a) DBRs based FP filter, and (b) ADMs-based FP filter. Reproduced from [27].

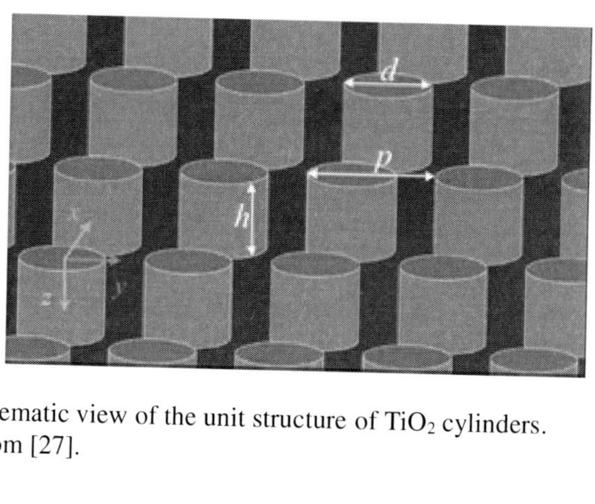

Figure 10. Schematic view of the unit structure of TiO_2 cylinders. Reproduced from [27].

Although DBR-based FP filters provide a simple way to build a tunable filter without modifying the cavity's physical length, its frequency response is limited to higher terahertz frequencies. On the other hand, the ADM reflectors are made up of TiO_2 cylinders that reflect more than 95.4 percent of light in a broadband range of 2.49–3.08 THz, which falls within the terahertz communication spectrum.

Table 1. Summary of FP filters

FP filter types	Composition	Tunable method	Reflection	Tunability
DBR Based	Si and SiO_2	Fill factor of MM	~99% between 166 – 230 THz	Up to 750 THz
ADMs Based	Benzocyclobutene (BCB) and TiO_2	Changing the cavity thickness.	~100% at 2.49 and 3.08 THz	2.49 – 3.08 THz

3.2. Graphene-Based Tunable Metamaterial Terahertz Filter

Graphene is made up of a monolayer of carbon atoms tightly arranged in a honeycomb (hexagonal) lattice structure. It has some interesting properties because of its lattice structure. Many research groups, particularly in THz, are interested in graphene due to its excellent properties like high thermal and electrical conductivity, hardness, resistance, etc. [28, 29]. The optical conductivity of graphene at low THz range follows a Drude-like behaviour and is heavily dependent on the fermi level of graphene, which can be altered using field effect. Based on these properties, many reconfigurable THz devices have been developed. At high fields and normal THz incidence, graphene shows a third-order non-linear response with magnitude greater than that achieved with infrared or visible spectrum.

A tunable graphene filter with a linear THz response is shown in Figure 11(a). The filter structure is consisted of metallic rings arranged periodically with a split between them. The split gap is composed of tunable graphene sheets, as shown in Figure 11. (b). The tunability, or resonant frequency, can be varied by changing the conductivity and the stacked layers of graphene. Also, the tuning factor can be controlled by changing the radius (inner and outer) of the metal rings. The designed filter offers tunability by changing graphene's conductivity by 40%. The tuning range is 0.2 THz to 0.12 THz [30].

The usage of a periodic arrangement of graphene causes the resonance frequency to shift despite modulation at the resonance frequency. Furthermore, the changes in geometry through chemical vapour deposition (CVD) makes a significant change in frequency (i.e., from 0.2 THz to 0.12 THz). Thus, the filter can be scaled to any frequency in the THz range by simply changing its geometry.

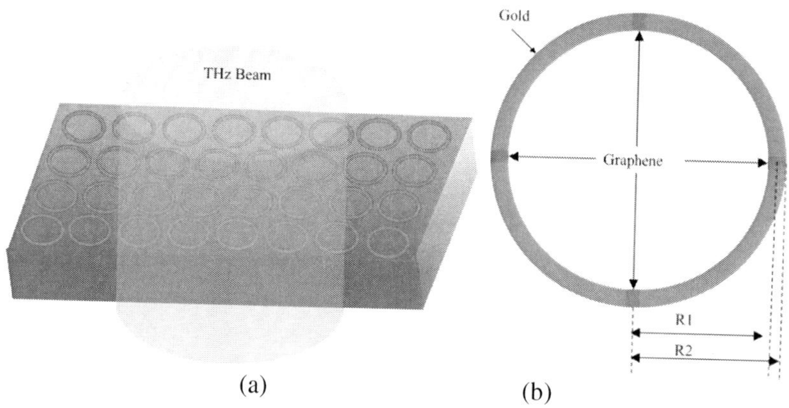

Figure 11. (a) THz MMF based on graphene, (b) Unit cell with inner and outer radius. Reproduced from [30].

Table 2. Summary of Graphene based filters

Filter type	Composition	Tunable method	Active area ratio	Tunability
Graphene Based	Glass, Graphene and Gold	Changing the Conductivity of Graphene	0.032	0.12 – 0.2 THz

The performance of these devices is greatly influenced by the switching speed limited by RC time constant limitations. Therefore, to increase the switching speed, the active area is made smaller as they are inversely proportional to each other. Further, it is inferred that the THz filter resonance can be adjusted by tuning the graphene conductivity by altering the number of graphene layers.

3.3. Electromagnetically Induced Transparency-Based Terahertz Filter

Another interesting phenomenon that has gained popularity due to its ability to remove absorption in a narrow band is used in the design of metamaterial THz filter. The above effect, caused by quantum interference, is known as electromagnetically induced transparency (EIT), and it can manipulate the THz spectrum [31-33]. In non-quantum systems, this quantum effect is realized by coupling it with harmonic oscillators. In particular, the interference from the surface plasmon polariton (SPP) can be generated by coupling the bright and dark modes. Direct coupling with the incident electric field causes the bright mode, while the bright mode triggers the dark mode. Due to the properties mentioned above, EIT has emerged as a potential candidate in the slow light field and non-linear sensors. The EIT plasmonic behavior is easy to realize using metamaterials as it is designed based on subwavelength dimensions. The EIT effect can be used to design a wide range of components.

Several research groups are actively involved in experimenting with EIT-like effects. The steady growth in studies such as the investigation of different EIT effects among gold, silver, and aluminum metamaterials by Hokari et al. [34] leads to the successful demonstration of the EIT-like effect with gold metamaterials containing a cut wire, and a wire pair corresponds to the bright and dark mode, respectively, by Nakamura et al. [35].

Based on the EIT effect, an actively tunable THz MMF with MEMS for changing the resonant frequency was designed. The THz MMF consists of a movable bar and a fixed wire pair that serve as bright mode and dark mode resonators, respectively. As a result, the THz frequency can be manipulated by bringing the movable bar close to the fixed bar using the MEMS actuator driven by the applied voltage. The metamaterial can control the EIT effect. Therefore, the THz waves can be manipulated by simply changing the distance between the movable and fixed bars. Moreover, the MEMS actuation system used in the THz MMF (shown in Figure 12) requires a small driving voltage and delivers a high response, making it an unbeatable candidate for EIT-based THz MMF.

The EIT-based THz filters offer great advantages by taking MEMS technology in miniaturization and integration into real optical systems. Furthermore, the filter is polarization-dependent, making it a promising choice for active modulation, switching, memory, and other THz wave spectrum applications.

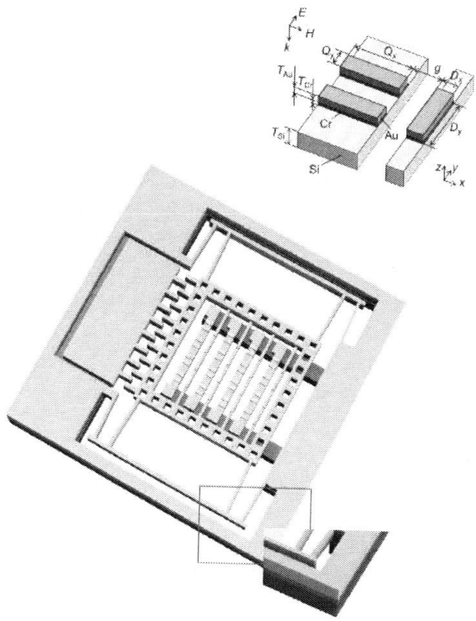

Figure 12. (a) MEMS-based THz metamaterial filter and (b) Unit cell of the EIT based THz MMF. Reproduced from [36].

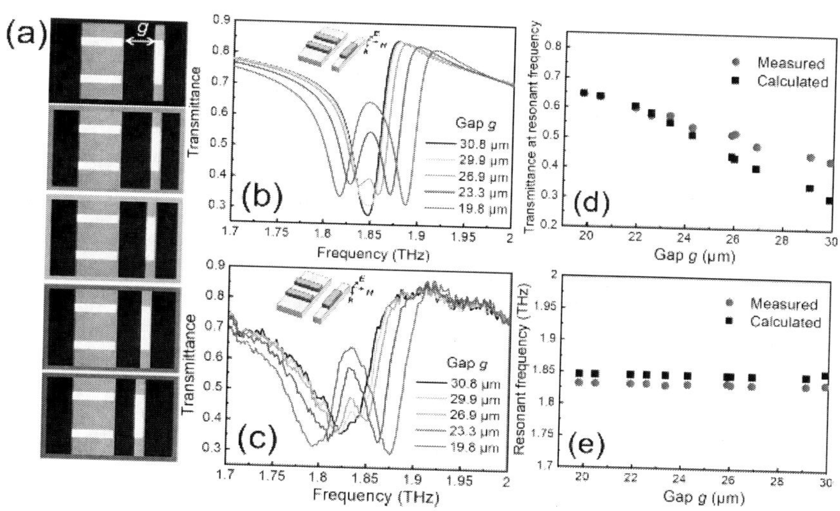

Figure 13. (a) Unit cell with a resonator gap (30.8 to 19.8 μm), (b) Calculated transmittance response, (c) measured transmittance response, (d) The resonant frequency's transmittance, and (e) Resonant frequency between the measured (red dots) and calculated (black squares) results. Reproduced from [36].

Table 3. Summary of EIT based filters

Filter type	Composition	Tunable method	Transmittance	Tunability
EIT Based	Silicon Chromium and Gold	Changing the distance (MEMS)	64.5%	1.0–1.832

3.4. Low Pass Terahertz Filter

SRR is used to construct a low pass filter (LPF) that operates in the THz spectrum [37]. The filter is composed of I-shaped and square-shaped copper rings, and the overall size is 35.48 × 21.15 μm^2. The LPF offers a low insertion loss at the cut-off frequency, (F_c) = 6.4 THz. The filter design can be improved by independently including a unit cell of SRR to the LPF.

The filter works by storing electromagnetic energy in a medium known as a resonator. The filter is based on the concept of SRR that is explained above. The split gap stores the electric field by providing larger capacitance. The electric field induces a magnetic field, storing the inductance across the two rings. The outer ring of the filter is designed as a square structure, while the inner ring is designed as an I structure. The resonant frequency of the filter can be changed by varying different parameters such as loop length, split gap distance, design material, etc. The SRR-based filter offers a simple way to design filters in the THz range. If the gap between the loops is kept small, it also provides miniaturization by providing better coupling. The overall gain is increased, and the losses are decreased by making the filter dimensions smaller than the resonant wavelength. The filter structure is described as a parallel combination of capacitors and inductors that create SRR by electromagnetic coupling between the two concentric rings [38, 39]. It is observed that the SRR resonance has negative permittivity, which is responsible for clean stopband rejection [40]. The equivalent circuit of the filter is shown in Figure 14. (b). The overall impedance of the *LC* circuit is shown in Eq. (2).

$$Z_1 = \frac{1}{j\omega C_1 + \frac{1}{j\omega L_1}} \tag{2}$$

The impedance of the capacitor is given by

$$Z_2 = \frac{1}{j\omega C_2} \tag{3}$$

The resonant frequency of the parallel branch at $Z_1 + Z_2 = 0$ is given as

$$f_s = \frac{1}{2\pi\sqrt{L_1(C_1+C_2)}} \tag{4}$$

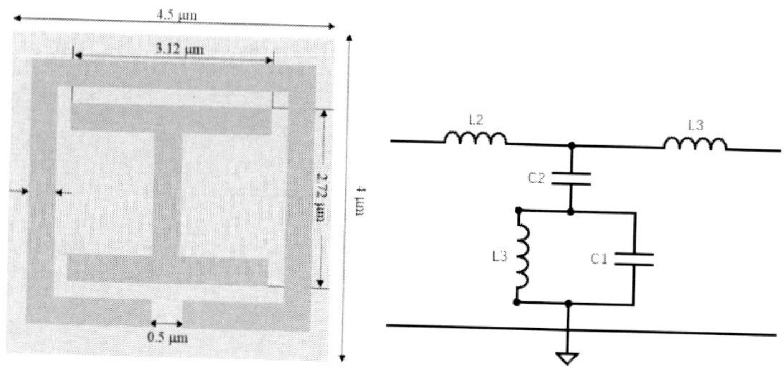

Figure 14. Split-ring resonator (a) Unit cell structure and (b) Equivalent circuit of the unit cell. Reproduced from [37].

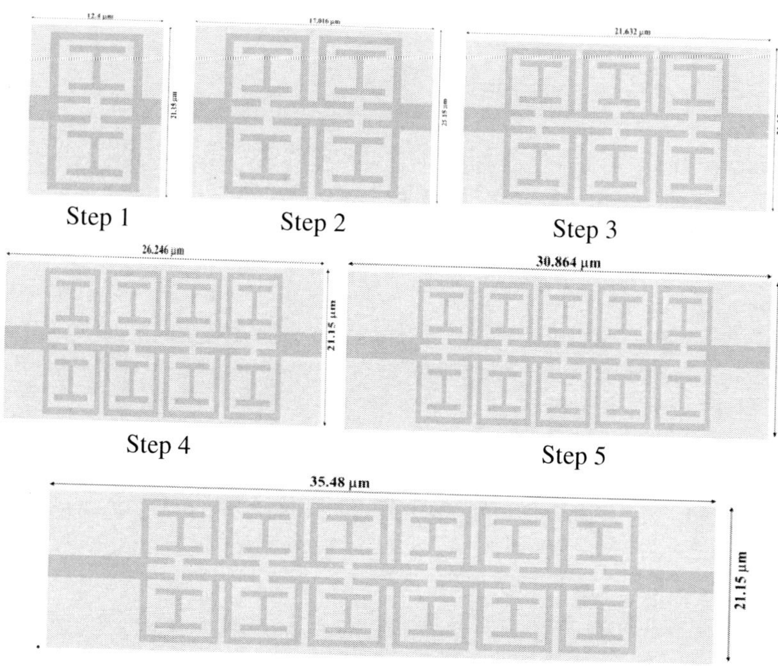

Figure 15. Design procedure of the THz LPF. Reproduced from [37].

The parallel circuit has a zero position at resonance, which is visible in the transmission function. The step-by-step design procedure of the SRR-based THz LPF is given in Figure 15.

Table 4. Summary of LPF

Filter type	**Composition**	**Method**	**Return Loss**	**Cut off frequency**
Low Pass	Polyimide and Copper	Periodic arrangement of SRR	-20 dB	6.4 THz

This filter can be used in submillimetre applications by making it as a reconfigurable filter with the help of tunable components like PIN diode and varactor diode. It can be easily combined with active and passive elements and has a low fabrication cost, as well as a wide rejection band and compact dimensions. Despite the filter's many benefits, one of its most significant flaws is its rejection band, which should be three times the bandpass in general, and this component of the filter has to be addressed.

3.5. Bandpass Terahertz Filter

Different techniques can be used to develop bandpass filters (BPF) in the THz regime. In this section, two types of BPF are explained. A metal-mesh BPF and a multilayer metamaterial-based BPF. The metal mesh BPF consists of a metal mess made up of copper film (12 μm thick) that operates in the frequency range of 0.5 – 2.5 THz. The frequency bandwidth is improved by placing two identical filters apart as shown in Figure 16. This arrangement results in a narrower frequency width at half maximum (FWHM). The effect of an incident wave at different angles has also been investigated.

Based on the studies, a new type of BPF called a spider filter (metal-mesh) was designed [41]. The designed BPF shown in the figure has been proven efficient in terms of reduced output noise and better transmission coefficient.

The frequency peak of the filter can be changed by varying its length, width, and structure (Figure 15, Eq. 5) [42].

$$\lambda_R = 2(g - 2a - b) \tag{5}$$

The substrate and the cylindrical column make up the second THz BPF based on multilayer metamaterials. Both the substrate and the cylindrical column are made of SiO_2. A thin layer of gold (0.2 μm) is used on both the cylindrical column and the substrate. The bandwidth of the THz BPF can be increased by increasing the number of cylinders in the unit cell. The filter is easy to design and can be made tunable by changing its dimensions, allowing for a wide range of wideband tunable THz filter implementations. The designed THz BPF has a bandwidth of 1 THz and transmittance of nearly 93% [43].

The designed filter is analyzed using the finite integration technique (FIT) as it requires less memory than the finite element method (FEM) [44]. In the simulation section, the systematic analysing method, i.e., the unit cell, is used. The boundary conditions are applied periodically along the *x*- and *y*-axes and the *z*-axis is left open [45].

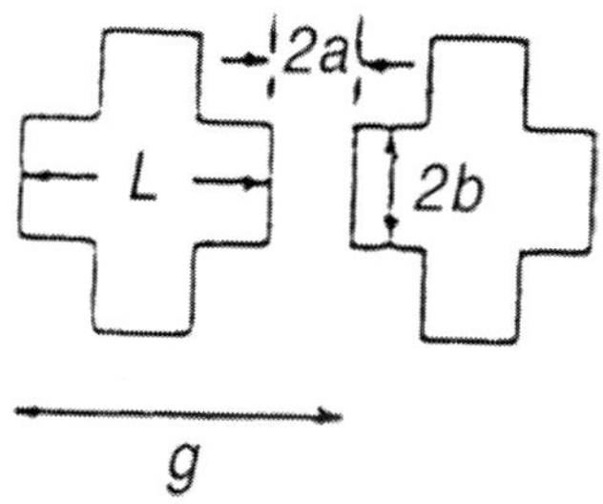

Figure 16. Plus-shaped BPF. Reproduced from [41].

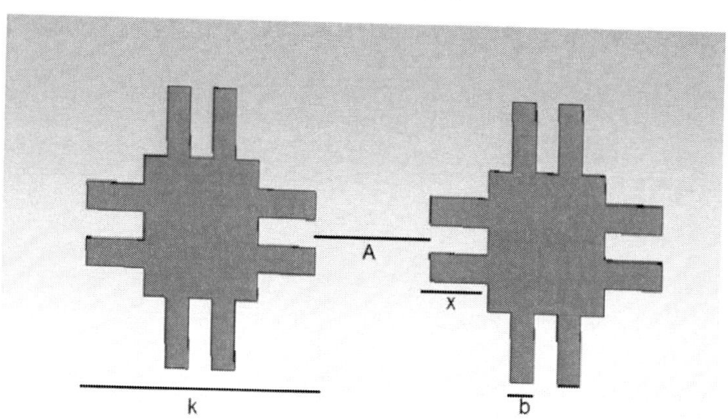

Figure 17. Spider shaped BPF. Reproduced from [41].

Table 5. Summary of BPF

Bandpass filter types	Composition	Design	Transmission Coefficient	Bandwidth
Metal mesh based	Air and Copper	Spider	~100%	~2 THz
Multilayer metamaterial based	SiO_2 and Gold	Cylindrical arrays.	~93%	~ 1 THz

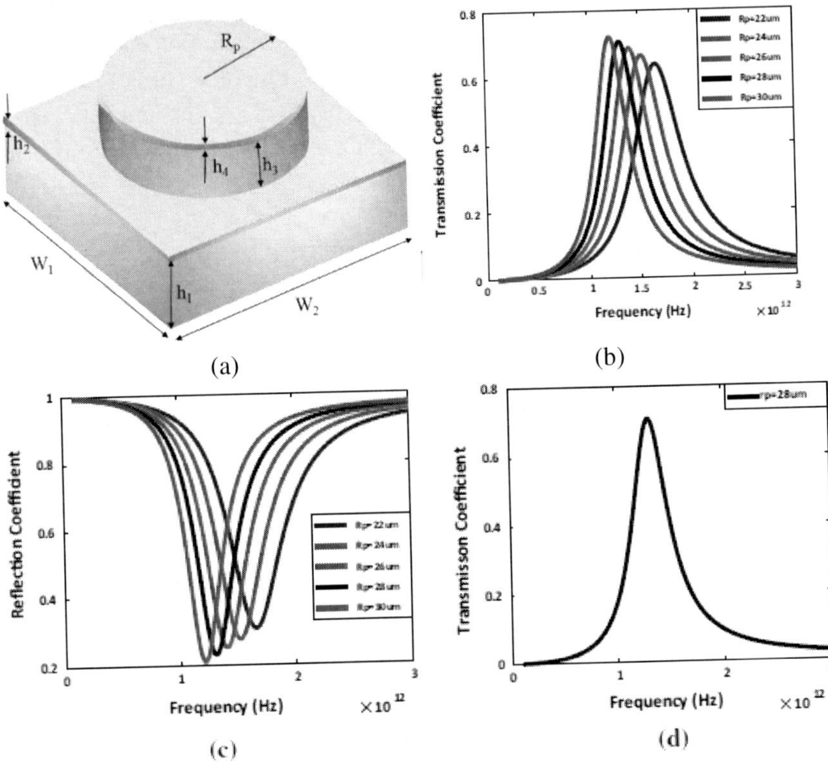

Figure 18. Simulated results for cylinder type THz BPF: (a) Unit cell, (b) Transmission coefficients, (c) Reflection coefficient for different radii (from 22 μm (blue) to 30 μm (magenta)), and (d) Frequency response (passing band) of the filter with radius of the particle $Rp = 28$ μm ($f_0 = 0.9$ THz, $T = 0.7$, $BW = 0.2$ THz). Reproduced from [43].

The results achieved substantiate that in the range of 0.1–3 THz, the performance of spider bandpass filter in terms of transmission coefficient has proved to be good along with much lower noise than the other bandpass filters ever made. However, the multilayer-based filter provides better tunability by increasing the metamaterial layers.

3.6. Comparison of Different MM-based THz Filters

Table 6. Comparison of different types of THz MMF

Ref.	Filter Type	Substrate	Tunability (THz)	Metal strip	Frequency (THz)
[20]	Fabry-Perot	Benzocyclobutene (BCB)	2.49 – 3.08 THz	TiO_2	2.78
[23]	Graphene-based Tunable THz	Glass	0.12 – 0.2	Graphene and Gold	0.2
[29]	EIT-based THz	Silicon	1.0–1.832	Chromium and Gold	1.832
[30]	Low pass THz	Polyimide	---	Copper	6.4
[35]	Band pass THz	Air	---	Copper	0.5–2.5

4. Potential Applications

Filters are used in almost every application, including imaging, communication, and spectroscopy [46]. Below are a few examples of potential applications that include filters.

4.1. Communication Systems

Terahertz communication is the most efficient way of utilizing the small range (up to few meters) with extremely high data rates (up to Tbps). There are several THz sources, such as interdigitated photoconductive antenna, spintronic materials, etc. which can generate broadband THz radiation [47-50]. Using such kind of sources, offering high data rates is often called "information shower" or "data shower." THz access points (APs) are deployed in areas such as railway stations, public buildings, shopping malls, and halls, where the human crowd is more. The AP can provide high data rates (up to several GBs), and the user can simply access it by just passing through it. Such data showers can be used to send software updates as well as other sorts of high-volume traffic, such as high-definition video, smoothly (e.g., a movie to watch in a bus, train etc.).

4.2. Miniaturization

THz opens up the possibility in realising micro- and nano-scale robots with the creation of micro-scale transceivers. A single nanorobot is basically incapable of performing any task alone due to its size and power consumption limitations. However, by connecting numerous nanorobots and making a network, they can perform various tasks that can benefit society in various ways, including environmental sensing and medicine. Furthermore, on-chip and chip-to-chip communications can only be enabled with very smaller transceivers (sub-mm scale), necessitating the usage of THz band radio. THz electronics based on graphene are now seen as a potential technique for creating huge multicore wireless networks-on-chip.

4.3. Biomedical Applications

In recent years, extensive research in the THz spectrum has resulted in the development of a number of applications. Since THz is non-ionising, it has been found to be a promising alternative to x-ray imaging. THz imaging can be used in biomedical applications as it produces images with high contrast, which aids in tissue classification and the detection of skin diseases.

4.4. Space Applications

THz photons are emitted during changes in atoms' molecular, rotational, or bending states [51]. THz spectroscopy also can observe physical parameters such as heat, pressure, etc., making it a suitable candidate for space-related applications like earth observation science, radio astronomy, and planetary or cometary science.

4.5. Security Applications

THz radiation can pass through many materials, including cloths, papers, plastics, etc., all of which are safe. Therefore, THz radiation can be used in airports to detect unauthorized items such as drugs, explosives, weapons, and so on [52]. The adhoc network can be formed in the military, consisting of numerous heterogeneous units like armed forces, armoured personnel carriers, tanks, etc. The fundamental advantage of the THz adhoc network over lower frequencies is the physical incapacity of any node outside the transmitter beam to eavesdrop or even notice the transmission. Therefore, the geometry of the network itself ensures better security along with the proper encryption schemes of the transmitted data.

4.6. Miscellaneous

Apart from the applications mentioned above, there are some others, such as THz sensing. THz-based sensors include soil, gas, medical, and so on. Parameters such as molecular density, water content, heat, pressure, and other physical parameters can be sensed using THz sensors. The major advantage of THz sensors is their ability to handle large amounts of data at high speeds in the Tbits/sec range. Therefore, THz sensors will be useful when continuous monitoring and high-speed processing are required.

Conclusion

THz MMFs, as well as their potential applications, are discussed in this book chapter. The difference between the THz MMFs is shown in table 1. It is clear that metamaterial THz filters are the right candidates for THz technologies, as they are used in almost every application, including imaging, communication, and spectroscopy. By inheriting metamaterial properties, the THz MMs can also be extended to design other components such as absorbers, modulators, and antennas.

References

[1] Pendry, John B., Anthony J. Holden, David J. Robbins, and W. J. Stewart. "Magnetism from conductors and enhanced nonlinear phenomena." *IEEE transactions on microwave theory and techniques* 47, no. 11 (1999): 2075-2084.

[2] Yen, Ta-Jen, W. J. Padilla, Nicholas Fang, D. C. Vier, D. R. Smith, J. B. Pendry, D. N. Basov, and Xiang Zhang. "Terahertz magnetic response from artificial materials." *Science* 303, no. 5663 (2004): 1494-1496.

[3] Engheta, Nader, and Richard W. Ziolkowski, eds. *Metamaterials: physics and engineering explorations*. John Wiley & Sons, 2006.

[4] Oueslati, Afef, Aymen Hlali, and Hassen Zairi. "Modeling of a metamaterial biosensor based on split ring resonators for cancer cells detection." *In 2021 18th International Multi-Conference on Systems, Signals & Devices (SSD)*, pp. 392-396. IEEE, 2021.

[5] Lim, C. S., M. H. Hong, Z. C. Chen, N. R. Han, B. Luk'Yanchuk, and T. C. Chong. "Hybrid metamaterial design and fabrication for terahertz resonance response enhancement." *Optics Express* 18, no. 12 (2010): 12421-12429.

[6] Suen, Jonathan Y., Kebin Fan, John Montoya, Christopher Bingham, Vincent Stenger, Sri Sriram, and Willie J. Padilla. "Multifunctional metamaterial pyroelectric infrared detectors." *Optica 4, no*. 2 (2017): 276-279.

[7] Tsakmakidis, Kosmas L., Allan D. Boardman, and Ortwin Hess. "'Trapped rainbow'storage of light in metamaterials." *Nature* 450, no. 7168 (2007): 397-401.

[8] Alu, Andrea, and Nader Engheta. "Dielectric sensing in ϵ-near-zero narrow waveguide channels." *Physical Review B* 78, no. 4 (2008): 045102.

[9] Chen, Hou-Tong, Willie J. Padilla, Joshua MO Zide, Arthur C. Gossard, Antoinette J. Taylor, and Richard D. Averitt. "Active terahertz metamaterial devices." *Nature* 444, no. 7119 (2006): 597-600.

[10] Wang, Qiugu, Depeng Mao, Peng Liu, Thomas Koschny, Costas M. Soukoulis, and Liang Dong. "NEMS-based infrared metamaterial via tuning nanocantilevers within complementary split ring resonators." *Journal of Microelectromechanical Systems* 26, no. 6 (2017): 1371-1380.

[11] Li, Yunjia, Dimitra Psychogiou, Stéphane Kühne, Jan Hesselbarth, Christian Hafner, and Christofer Hierold. "Large stroke staggered vertical comb-drive actuator for the application of a millimeter-wave tunable phase shifter." *Journal of microelectromechanical systems* 22, no. 4 (2013): 962-975.

[12] Xu, Tao, and Yu-Sheng Lin. "Tunable terahertz metamaterial using an electric split-ring resonator with polarization-sensitive characteristic." *Applied Sciences* 10, no. 13 (2020): 4660.

[13] Bi, Ke, Qingmin Wang, Jianchun Xu, Lihao Chen, Chuwen Lan, and Ming Lei. "All-dielectric metamaterial fabrication techniques." *Advanced Optical Materials* 9, no. 1 (2021): 2001474.

[14] Park, Sang-Gil, Kanghee Lee, Daehoon Han, Jaewook Ahn, and Ki-Hun Jeong. "Subwavelength silicon through-hole arrays as an all-dielectric broadband terahertz gradient index metamaterial." *Applied Physics Letters* 105, no. 9 (2014): 091101.

[15] Zhao, Yanjiao, Baiwei Li, Chuwen Lan, Ke Bi, and Zhaowei Qu. "Tunable silicon-based all-dielectric metamaterials with strontium titanate thin film in terahertz range." *Optics express* 25, no. 18 (2017): 22158-22163.
[16] Bi, Ke, and Jianchun Xu. "Tunable Metamaterials." *In Metamaterials*, pp. 83-115. CRC Press, 2021.
[17] Němec, H., P. Kužel, F. Kadlec, Christelle Kadlec, R. Yahiaoui, and Patrick Mounaix. "Tunable terahertz metamaterials with negative permeability." *Physical Review B* 79, no. 24 (2009): 241108.
[18] Huang, Kai, Junchen Liu, Lu Wang, Gang Chang, Ruyue Wang, Ming Lei, Yonggang Wang, and Yunbin He. "Mixed valence CoCuMnOx spinel nanoparticles by sacrificial template method with enhanced ORR performance." *Applied Surface Science* 487 (2019): 1145-1151.
[19] Zeng, Xinxi, Rong Wang, Xiaoqing Xi, Bo Li, and Ji Zhou. "3D direct writing of terahertz metamaterials based on TbFeO3 dielectric ceramics." *Applied Physics Letters* 113, no. 8 (2018): 081901.
[20] 20 Lu, Mingzhi, Wenzao Li, and Elliott R. Brown. "High-order THz bandpass filters achieved by multilayer complementary metamaterial structures." *In 35th International Conference on Infrared, Millimeter, and Terahertz Waves*, pp. 1-2. IEEE, 2010.
[21] Gat, Nahum. "Imaging spectroscopy using tunable filters: a review." *Wavelet Applications* VII 4056 (2000): 50-64.
[22] Al Sayem, Ayed, and Md Saifur Rahman. "Tunable Fabry Perot filter using all-dielectric multilayer anisotropic metamaterial in the mid-infrared frequency range." *In 2016 9th International Conference on Electrical and Computer Engineering (ICECE)*, pp. 361-364. IEEE, 2016.
[23] Correia, José Higino, Marian Bartek, and Reinoud F. Wolffenbuttel. "High-selectivity single-chip spectrometer in silicon for operation in visible part of the spectrum." *IEEE Transactions on Electron Devices* 47, no. 3 (2000): 553-559.
[24] Emadi, Arvin, Huaiwen Wu, Ger de Graaf, and Reinoud Wolffenbuttel. "Design and implementation of a sub-nm resolution microspectrometer based on a Linear-Variable Optical Filter." *Optics express* 20, no. 1 (2012): 489-507.
[25] O'Brien, Nada A., Charles A. Hulse, Donald M. Friedrich, Fred J. Van Milligen, Marc K. von Gunten, Frank Pfeifer, and Heinz W. Siesler. "Miniature near-infrared (NIR) spectrometer engine for handheld applications." *In Next-generation spectroscopic technologies* V, vol. 8374, pp. 31-38. SPIE, 2012.
[26] Xiao, Jing, Fuchuan Song, Kijeong Han, and Sang-Woo Seo. "Fabrication of CMOS-compatible optical filter arrays using gray-scale lithography." *Journal of Micromechanics and Microengineering* 22, no. 2 (2012): 025006.
[27] Ao, Tianhong, Xiangdong Xu, Yu Gu, Yadong Jiang, Xinrong Li, Yuxiang Lian, and Fu Wang. "A tunable Fabry–Perot filter (λ/18) based on all-dielectric metamaterials." *Optics Communications* 414 (2018): 160-165.
[28] Sensale-Rodriguez, Berardi, Rusen Yan, Lei Liu, Debdeep Jena, and Huili Grace Xing. "Graphene for reconfigurable terahertz optoelectronics." *Proceedings of the IEEE* 101, no. 7 (2013): 1705-1716.

[29] Sensale-Rodriguez, Berardi, Rusen Yan, Mingda Zhu, Debdeep Jena, Lei Liu, and Huili Grace Xing. "Efficient terahertz electro-absorption modulation employing graphene plasmonic structures." *Applied Physics Letters* 101, no. 26 (2012): 261115.

[30] Yang, Kai, Shuchang Liu, Sara Arezoomandan, Ajay Nahata, and Berardi Sensale-Rodriguez. "Graphene-based tunable metamaterial terahertz filters." *Applied Physics Letters* 105, no. 9 (2014): 093105.

[31] Harris, Stephen E., J. E. Field, and A. Imamoğlu. "Nonlinear optical processes using electromagnetically induced transparency." *Physical Review Letters* 64, no. 10 (1990): 1107.

[32] Fleischhauer, Michael, Atac Imamoglu, and Jonathan P. Marangos. "Electromagnetically induced transparency: Optics in coherent media." *Reviews of modern physics* 77, no. 2 (2005): 633.

[33] Boller, K-J., A. Imamoğlu, and Stephen E. Harris. "Observation of electromagnetically induced transparency." *Physical Review Letters* 66, no. 20 (1991): 2593.

[34] Hokari, Ryohei, Yoshiaki Kanamori, and Kazuhiro Hane. "Comparison of electromagnetically induced transparency between silver, gold, and aluminum metamaterials at visible wavelengths." *Optics express* 22, no. 3 (2014): 3526-3537.

[35] Nakamura, Kenta, Ryohei Hokari, Yoshiaki Kanamori, and Kazuhiro Hane. "Fabrication of electromagnetically induced transparency like metamaterials in THz region and evaluation of the transmittance characteristics." *IEEJ Transactions on Sensors and Micromachines* 135, no. 11 (2015): 454-459.

[36] Huang, Ying, Kenta Nakamura, Yuma Takida, Hiroaki Minamide, Kazuhiro Hane, and Yoshiaki Kanamori. "Actively tunable THz filter based on an electromagnetically induced transparency analog hybridized with a MEMS metamaterial." *Scientific reports* 10, no. 1 (2020): 1-10.

[37] Singh, Joy, Sudhanshu Kumar Jha, Vimlesh Singh, and Y. K. Awasthi. "Design of THz low pass filter using split-ring resonators." *Optik* 247 (2021): 167925.

[38] Thulaseedharan Rekha, Parambil Abdulla, Paruthikkal Mohammed Raphika, and Puthenveetil Muhammed Jasmine, "Compact Microstrip Lowpass Filter with Ultra-Wide Stopband Using Patch Resonators and Open Stubs," *Progress In Electromagnetics Research C*, Vol. 72, 15-28, 2017.

[39] Rekha, Thulaseedharan, Parambil Abdulla, Paruthikkal Mohammed Raphika, and Puthenveetil Muhammed Jasmine. "Compact microstrip lowpass filter with ultra-wide stopband using patch resonators and open stubs." *Progress in Electromagnetics Research C* 72 (2017): 15-28.

[40] Zeng, Hui-Yong, Guang-Ming Wang, Chen-Xin Zhang, and L. I. Zhu. "Compact microstrip low-pass filter using complementary split ring resonators with ultra-wide stopband and high selectivity." *Microwave and Optical Technology Letters* 52, no. 2 (2010): 430-433.

[41] Zeng, Hui-Yong, Guang-Ming Wang, Chen-Xin Zhang, and L. I. Zhu. "Compact microstrip low-pass filter using complementary split ring resonators with ultra-wide stopband and high selectivity." *Microwave and Optical Technology Letters* 52, no. 2 (2010): 430-433.

[42] Möller, K. D., J. B. Warren, J. B. Heaney, and C. Kotecki. "Cross-shaped bandpass filters for the near-and mid-infrared wavelength regions." *Applied optics* 35, no. 31 (1996): 6210-6215.

[43] Asl, A. Beheshti, A. Rostami, and I. S. Amiri. "Terahertz band pass filter design using multilayer metamaterials." *Optical and Quantum Electronics* 52, no. 3 (2020): 1-13.

[44] Rahimi, Zhabiz. "The finite integration technique (FIT) and the application in lithography simulations." *Friedrich-Alexander-Universitaet Erlangen-Nuernberg (Germany)*, 2011.

[45] Beheshti Asl, A., A. Rostami, and I. S. Amiri. "Radiation pattern direction control in nano-antenna (tunable nano-antenna)." *Optical and Quantum Electronics* 51, no. 11 (2019): 1-13.

[46] Choudhury, Balamati, Sanjana Bisoyi, Pavani Vijay Reddy, S. Manjula, and R. M. Jha. "Emerging trends in terahertz metamaterial applications." *Comput. Mater. Continua* 39 (2014): 179-215.

[47] Boby, E., Prajapati, J., Rathinasamy, V., Mukherjee, S. and Mondal, S., 2022. "Parametric Investigation of Interdigitated Photoconductive Antenna for Efficient Terahertz Applications." *Arabian Journal for Science and Engineering*, *47*(3), pp. 3597-3609.

[48] Boby, E. N. F., Prajapati, J., Rathinasamy, V., Rao, T. R. and Mondal, S., 2021. "6G and beyond: Investigation of broadband terahertz interdigitated photoconductive antenna by exploiting laser parameters." *Microwave and Optical Technology Letters.*

[49] Rathinasamy, V., Thipparaju, R. R., Edwin, N. F. B. and Mondal, S., 2021. "Interdigitated photoconductive terahertz antenna for future wireless communications."*Microwave and Optical Technology Letters.*

[50] Bull, C., Hewett, S. M., Ji, R., Lin, C. H., Thomson, T., Graham, D. M. and Nutter, P. W., 2021. "Spintronic terahertz emitters: Status and prospects from a materials perspective." *APL Materials*, *9*(9), p. 090701.

[51] de Maagt, Peter. "Terahertz technology for space and earth applications." *In 2007 International workshop on Antenna Technology: Small and Smart Antennas Metamaterials and Applications*, pp. 111-115. IEEE, 2007.

[52] Hashimshony, D., A. Zigler, and K. Papadopoulos. "Conversion of electrostatic to electromagnetic waves by superluminous ionization fronts." *Physical review letters* 86, no. 13 (2001): 2806.

Chapter 7

A Temperature Tunable Terahertz Metamaterial Absorber Based on InSb Substrate

Snehashis Basu[1], Ayesha Mohanty[1], Omprakash Acharya[1], Susanta Kumar Mohapatra[1], Bhargav Appasani[1] and Mohammad S. Khan[2,*]

[1]School of Electronics, Kalinga Institute of Industrial Technology, Bhubaneshwar, India
[2]Department of Computer & Information Sciences, East Tennessee State University, Johnson City, TN, USA

Abstract

A temperature tunable terahertz (THz) metamaterial absorber is proposed in this chapter for sensing related applications. The proposed design showed the capability of sensing variation in temperature of the surrounding medium. The design has a circular ring resonator (CRR) made out of gold laid over an Indium Antimonide (InSb) substrate. The sensor is found to be capable of achieving an absorption of 85% and above for different resonance frequencies. A complete parametric analysis was carried out on the design to justify the design dimensions. The chapter explains in detail about the physics beind the shift in resonance frequencies with the temperature. The temperature was varied between 285K and 310 K and its sensitivity was calculated to be 960 MHz/K. Hence, the proposed design can effectively be used as temperature sensor for biomedical sensing, embedded sensor related applications.

---

* Corresponding Author's Email: adhoc.khan@gmail.com.

In: Metamaterials for Microwave and Terahertz Applications
Editors: B. Appasani, O. Prakash Acharya, A. Vidyakant Jha et al.
ISBN: 979-8-88697-330-3

Keywords: terahertz metamaterial absorber, temperature sensor, indium antimonide, ring resonator

1. Introduction

In recent decades, condensed matter physics has grown at a breakneck speed, largely due to concurrent advances in processing power, material manufacturing, crystal growth, and experimental optical methods, among other things. Scientists and engineers can now conceive, design, and optimize a metamaterial, and its response before it ever leaves the drawing board, thanks to advances in computational power and simulation algorithms. Progress in nanoscale fabrication techniques has allowed designers to build increasingly complex and tiny structures with high fidelity, while advances in the fabrication of complex solid-state systems have allowed designers to integrate a wide range of complex constituent materials into metamaterial devices. Finally, new techniques have enabled the study of metamaterials in ever-widening time scales and frequency regimes, particularly in the terahertz regime. This convergence of development is most evident in the relatively new subject of terahertz metamaterials. Metamaterials are man-made materials with permittivity (ε) and permeability (μ) that are both negative for a given excitation wave frequency. Metamaterials have a negative refraction index as a result. These materials are also double-negative (DNG) or negative-index metamaterials (NIMs). Victor Veselago released his work on the issue in Russian in 1967, titled The Electrodynamics of Substances with Simultaneously Negative Values of ε and μ, which was later translated into English in 1968 [1]. However, the technology required to create such materials did not exist at the time. David Smith and Sir John Pendry found how to create metamaterials in the year 1999 physically. This resulted in a whole new field of study. Sir John Pendry demonstrated negative permittivity in a three-dimensional series of thin wires in 1999 [2]. Shortly after, the split ring resonators (SRRs) made up of copper with effective negative permeability were constructed. The two were combined in 2000 by Smith et al., to create the first double-negative metamaterial for a frequency band in the GHz range [3]. Synthetic structural materials with unique electromagnetic properties are known as electromagnetic metamaterials [4]. Electromagnetic metamaterials have various applications in electromagnetism, optics, and materials science [5-7]. The structural parameters of the unit cell and material qualities of substrates are strongly correlated with the resonant frequency and resonant-type electromagnetic metamaterials strength. Adjusting ambient physical and/or chemical factors can tune such structural characteristics and/or dielectric qualities.

The recent growth of communication has resulted in significant electromagnetic (EM) interference, which pollutes the environment. As a result, the normal conduct of technologies tempers the quality of humans. As a result, there is a growing need for materials and processes that might help alleviate the impacts of electromagnetic radiation. The advent of an absorber into the realm of communication has met this desire. The energy of an incoming particle is absorbed by an absorber, a block of material. Absorber research is getting popular due to its prospective uses in solar cells with great efficiency, sensors, and stealth technologies. Military uses or stealth technologies are the most prevalent among the numerous lucrative applications. Stealth technology aims to hide a plane from radar by coating it with radar-absorbing materials. The radar waves that reflect off the surface of an airplane can be reduced or blocked by the absorber employed in this technique. Dallenbach, Salisbury, and Jaumann absorbers are some famous examples of EM absorbers [8-11].

On the other hand, traditional absorbers have a big bulk and sophisticated design that makes them unsuitable for many practical uses [12-13]. As a result, research has switched to fulfilling the requirement for small, basic absorbers known as metamaterial absorbers (MMA). Metamaterial-based absorbers (MMAs) have received much interest because of their maximum absorption, thin sheets, and low density. An incident EM wave can be trapped or absorbed by MMAs in specified areas of the device, and the EM can subsequently be converted to heat. Since Landy revealed the ideal MMA in 2008 [14], metamaterial absorber MMA has been one of the most promising topics for researchers in recent years. The structures' remarkable features led to their quick adoption in a broad range of applications, from microwave to optical [15-18]. Between microwave and optical frequencies lies the terahertz (THz) frequency spectrum. The "terahertz gap" refers to a vacuum in the frequency spectrum that has yet to be thoroughly investigated. However, several new applications, such as terahertz imaging and terahertz sensing, have been created in recent decades. As these applications have grown in popularity, research into the creation of terahertz metamaterial absorbers (TMAs) has gained traction. The unit cells of a TMA are 2-dimensional or 3-dimensional periodic patterns. TMAs should be able to absorb terahertz electromagnetic radiation independent of polarization or incidence angle to be ideal. When T. J. Yen and his colleagues constructed and tested THz magnetic response using metamaterials, they laid the groundwork for TMA [19]. TMAs cause the absorption spectra to develop by lowering the plasma frequency to the predicted operating frequency. Tunable TMAs are a kind of TMA that may have their absorption bands altered by incorporating external stimuli like mechanical force, temperature, and voltage [20–22].

As a result, novel TMA-based sensing technologies can be created. Metamaterial-inspired detection has various advantages compared to standard sensing methods, including high precision, operation without labels, and the capacity to wirelessly operate across large distances. Explorations of electromagnetic metamaterials-based sensing mechanisms, sensing technology, and device engineering are well-developed, and their implementation approaches in chemical detection, biosensing, environmental sensing, and upcoming Internet of Things (IoT) technologies have been extensively investigated in recent years [23–25]. Temperature adjustable TMAs are produced by using external stimuli such as temperature for modifying the absorption spectrum, and they have a variety of applications in the construction of terahertz temperature sensors [26]. Broadband absorption features have been built into single-band temperature TMAs. InSb or VO2 is the temperature-sensitive dielectric spacer employed in creating these devices, and its permittivity changes with temperature, resulting in spectrum shifting [27]. Because of the above-mentioned designs' single-band absorption properties, researchers focused on developing multi-band temperature adjustable TMAs. Temperature-sensitive materials, such as InSb, are commonly employed because their permittivity changes with temperature, causing spectra to shift.

Temperature sensing has been a key research and application in metamaterial based sensing. This is owing to the extreme temperature sensitivity of certain of the base material and/or sub-wavelength structures used. A unique temperature-tunable metamaterial absorber is examined in this chapter.

2. Design Structure

To design the metamaterial absorber structure for temperature sensing, a unit cell is considered with dimensions as follows, U = 200 μm, I = 25 μm, O = 75 μm, t = 2 μm, h = 10 μm and b = 0.2 μm. The schematic diagram of the novel design comprising top and side views is depicted in Figure 1.

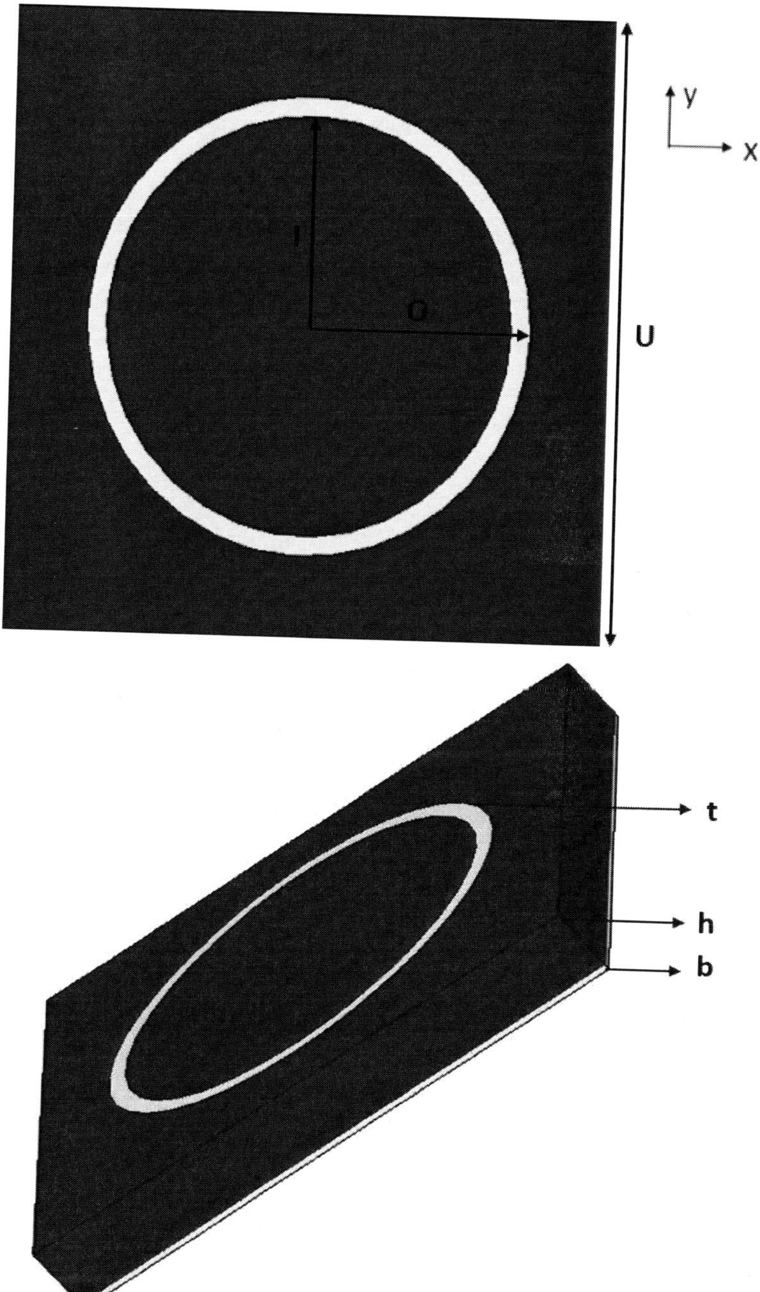

Figure 1. Schematic of MMA based temperature sensor.

The temperature controllable unit cell structure of the metamaterial absorber comprises of 3 layers. Normally metamaterial is built in a 3D structure; the first layer is the top metallic layer on which patches are made. A dielectric spacer is placed in the middle layer, and the bottom layer has a metallic ground plane. In this proposed structure, gold having a conductivity of $\sigma = 4.07 \times 10^7$ S/m has been taken as the top layer metal and bottom layer. The substrate chosen for the design of the MMA is InSb.

Drude's model guides the change in the permittivity of substrate material, which is supplied by Eq (1). The equation applies to the temperature between 260 K and 320 K; the frequency is kept in the range of 0.1 - 2.2 THz [28].

$$\varepsilon_r(\omega) = \varepsilon_\infty - \frac{\omega_P^2}{\omega^2 + j\omega\gamma} \tag{1}$$

where, ω is operating frequency in rad/sec, ω_P is plasma frequency, γ is damping constant, and ε_∞ is high frequency permittivity ($\varepsilon_\infty = 15.6$ for InSb)

$$\omega_P = \sqrt{Ne^2/\varepsilon_0 m^*} \tag{2}$$

$$\gamma = \frac{e}{m^* \mu} \tag{3}$$

where, N is the intrinsic density of carriers, e is the charge of an electron, ε_0 is the permittivity of free space, m^* is the effective mass of free carriers, $m^* = 0.015$ m, and μ is electron mobility.

It has been discovered that the mobility of electrons varies little with temperatures ranging from 260K and 320K. Hence, the value of γ is assumed to be constant and taken as 0.1π THz.

$$N = 5.76 \times 10^{20} T^{1.5} e^{-\left(\frac{0.26}{2k_B T}\right)} \tag{4}$$

where, k_B - Boltzmann's constant, and T is the temperature in Kelvin.

The relative permittivity at different temperatures was calculated using MATLAB software. The suggested absorber device was designed and simulated using CST Microwave Studio's High Field Simulation Software (HFSS). The Finite Element Method (FEM), a numerical method used in HFSS, is used to model the recommended design. A structure is split into multiple little substructures called finite elements in this FEM. The tetrahedral finite element mesh pattern is used to insert these finite elements. For the entire structure, an iterative field solution is determined. Because the fields are interdependent, Maxwell's equations are applied across inter-element boundaries, resulting in a structural solution. After getting the field solution, the generalized S-parameter of the structure is determined. The S-parameter is then used to calculate absorption. Electric and magnetic fields, instead of voltages and currents, are used to build structures in HFSS. To demonstrate the behavior of the EM field of the whole structure in HFSS, periodic boundary conditions are applied to the solution's outer surface. The proposed model is closed, meaning that no energy exits the system except through the excitation port.

The approach uses a planar EM beam to emit the desired structure. The periodic arrangement of the MMA is then revealed by periodic boundary conditions put along the z axis. For excitation, the absorber's ports are assigned in the z- direction. The equation (5) below is used to compute the absorption (A).

$$A = 1 - R - T = 1 - |s_{11}|^2 - |s_{12}|^2 \tag{5}$$

where A is for absorption, T is for transmission, and R is for reflection

Because the metallic bottom layer completely blocks the input beam's transmission, the absorption coefficient (A) in the preceding equation is exclusively dependent on the reflection coefficient (R).

3. Simulations Results of a Temperature Tunable MMA

The finite-element approach (FEM) is used to model the proposed metamaterial absorber. On four sides of the unit cell, excitation ports are given in the z-direction, and along the Z-plane, boundary conditions are supplied. The absorption spectra for the temperature tunable metamaterial absorber are shown in Figure 2. The metamaterial absorber is simulated using the software for the temperatures 285 K, 290 K, 295 K, 300 K, 305 K and 310 K. This tri-layer metamaterial absorber structure can be used as a temperature sensor as it can be tuned by varying the temperature to the presented MMA. With the temperature change, absorption changes accordingly. The shift in the absorption band is seen in Figure 2.

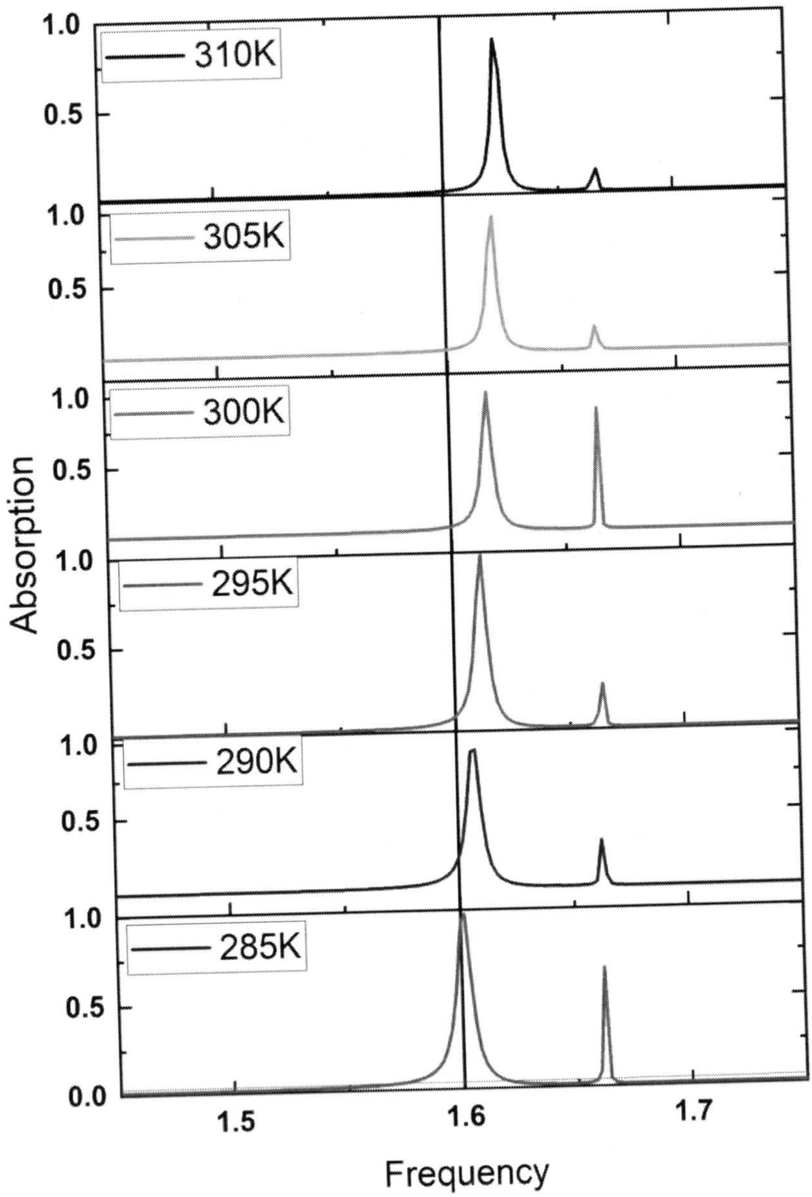

Figure 2. Absorption spectra of the temperature tunable MMA.

When $T = 285K$, the peak absorption was found to be 94.7% at 1.601 THz. When the temperature was increased to 290K, the absorption peak shifted to

1.608 THz, with the peak absorption of 92.16%. At 295K, the resonance peak shifted to 1.612 THz, with a peak absorption of 98.01%. When the temperature is further increased to 300 K, the resonance peak shifted to 1.616 THz, with a peak absorption of 99.01%. When the temperature was increased to 305K, the peak occurred at 1.62 THz, with an absorption of 92.97%. Finally, at 310K, the resonance occurred at 1.626 THz, with peak absorption of 86.27%. The resonance frequency at different temperatures is plotted in Figure 3.

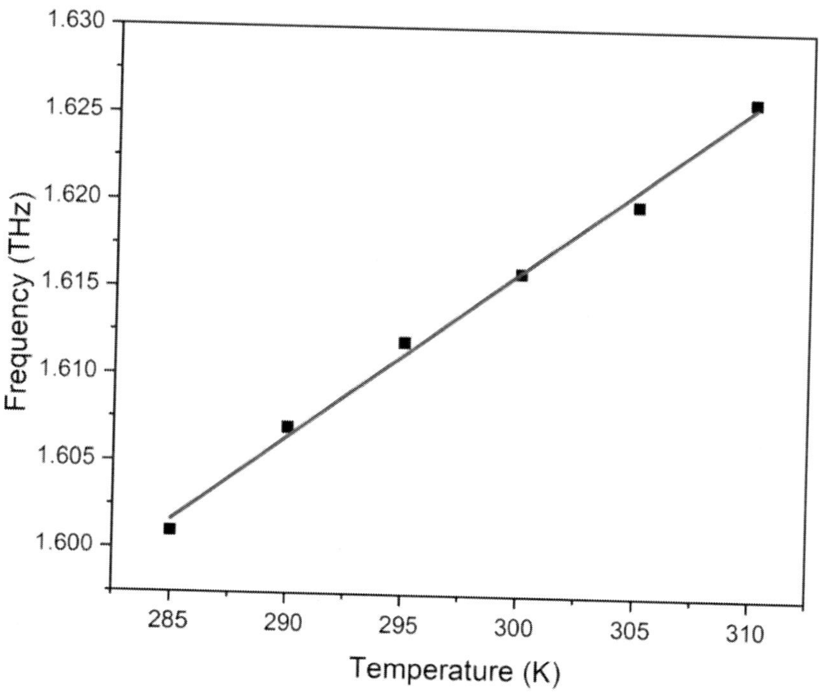

Figure 3. Resonance frequency vs. temperature plot.

The sensitivity of the sensor can be calculated as the ratio of shift in the resonance frequency per unit change in the tempurature, which is found to 960 MHz/K. This value is sufficient to detect the changes in the temperature of the surrounding medium. The absorption mechanism is studied further by using CST Microwave Studio's surface current simulator. The mechanism is investigated by current distribution plots for the main resonance at 300K.

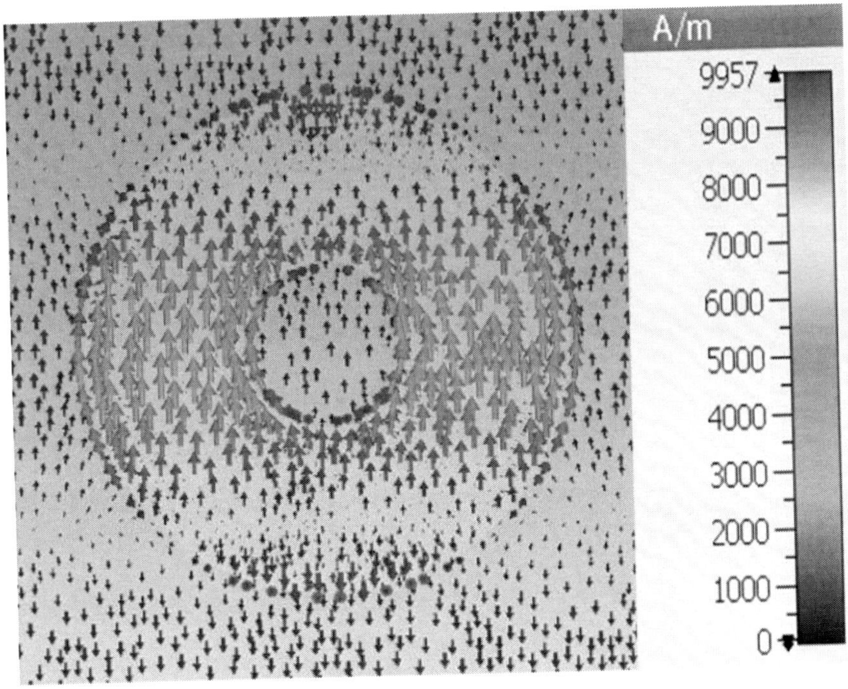

Figure 4. Current distribution of 300 K at 1.616 THz.

The current distribution for resonant peak at 1.616 THz is current concentrated at the center of the ring resonator. The oscillating current results in a magnetic resonance, leading to absorption of electromagnetic energy.

Conclusion

From this chapter, it is concluded that metamaterial absorbers can be utilized as temperature tunable absorbers. This novel absorber can be tuned by changing the plasma frequency of the dielectric spacer. This chapter investigates about the temperature tunable absorber by using a novel design consisting of a complete outer square ring and an inner split ring. The proposed structure yields a peak absorption of 99% at room temperature, i.e., at 300K. At other temperatures, resonant frequencies, the peak asborptions were found to be greater than 85%. The sensitivity of the sensor was fornd to 960 MHz/K, which is significant and can find applications in detecting changes at room temperature.

References

[1] Veselago V. G., "The electrodynamics of substances with simultaneously negative values of ε and μ," *Sov. Phys. Uspekhi*, vol. 10, no. 4, pp. 509–514, 1968. [Usp. Fiz. Nauk, vol. 92, pp. 517–526, 1967].

[2] Pendry J. B., A. J. Holden, D. J. Robbins, and W.J. Stewart, "Magnetism from Conductors and Enhanced Nonlinear Phenomena," *IEEE Transactions on Microwave Theory and Techniques*, Vol. 47, No. 11, pp. 2075-2084, 1999.

[3] Smith D. R., W. J. Padilla, D. C. Vier, S. C. Nemat-Nasser and S. Schultz, "Composite medium with simultaneously negative permeability and permittivity," *Physical Review Letters*, Vol. 84, No. 18, 4184–4187, 2000.

[4] Cui T. J., D. R. Smith, Ruopeng Liu, *"Metamaterials: Theory, Design and Applications.* New York", NY: Spring-Verlag, 2010.

[5] Xu, He-Xiu, Shulin Sun, Shiwei Tang, Shaojie Ma, Qiong He, Guang-Ming Wang, Tong Cai, Hai-Peng Li, and Lei Zhou. "Dynamical control on helicity of electromagnetic waves by tunable metasurfaces", *Sci. Rep.*, vol. 6, pp. 1-10, 2016.

[6] Xu H X, S Ma, W Luo, T Cai, S Sun, Q He. "Aberration-free and functionality-switchable meta -lenses based on tunable metasurfaces", *Appl. Phys. Lett.*, vol. 109, pp. 193506, 2016.

[7] Ren X, R Das, P Tran, T D Ngo, Y M Xie, " Auxetic metamaterials and structures: a review," *SmartMat Struct.*, vol. 27, pp. 023001, 2018.

[8] Munk B. A., *Frequency selective surfaces: theory and design.* John Wiley & Sons, 2005.

[9] Salisbury W. W., *"Absorbent body for electromagnetic waves."* Google Patents, 1952.

[10] Ruck G. T., *Radar cross section handbook, vol. 2.* Plenum Publishing Corporation, 1970.

[11] Chambers B. and A. Tennant, "Active Dallenbach radar absorber," in *2006 IEEE Antennas and Propagation Society International Symposium*, 2006, pp. 381–384.

[12] Engheta N., *"Metamaterials in Microwaves and Optics : A Review,"* pp. 387–389, 2006.

[13] Shamonina E. and L. Solymar, "Metamaterials : How the subject started," *metamaterials* 1, vol. 1, pp. 12–18, 2007.

[14] Landy N. I., S. Sajuyigbe, J. J. Mock, D. R. Smith, and W. J. Padilla, "Perfect Metamaterial Absorber," *Phys. Rev. Lett.*, vol. 100, no. 20, pp. 207402-1–4, May 2008.

[15] Zhang W. Feng Lan, Jinyang Xuan, Pinanki Mazumder, Mahdi Aghadjani, Ziqiang Yang., Lin Men. "Ultrasensitive dual-band terahertz sensing with metamaterial perfect absorber," *IEEE MTT-S International Microwave Workshop Series on Advanced Materials and Processes*, no. September, pp. 1–3, 2017.

[16] Bakshi S. C., D. Mitra, and L. Minz, "A Compact Design of Multiband Terahertz Metamaterial Absorber with Frequency and Polarization Tunability," *Plasmonics*, pp. 1–10, 2018.

[17] Wang B. X., X. Zhai, G. Z. Wang, W. Q. Huang, and L. L. Wang, "A novel dual-band terahertz metamaterial absorber for a sensor application," *Journal of Applied Physics*, vol. 117, no. 1, 2015.

[18] Li D., H. Huang, H. Xia, J. Zeng, H. Li, and D. Xie, "Temperature-dependent tunable terahertz metamaterial absorber for the application of light modulator," *Results in Physics*, vol. 11, pp. 659–664, 2018.

[19] Yen T. J., W. J. Padillan, Fang D. C., Vierd. R., Smith J. B., Pendry D. N. Basovand X. Zhang. "Terahertz Magnetic Response from Artificial Materials," *Science*, vol. 303, pp.1494-1496, 2004.

[20] He X Y, X Zhong, F T Lin, W Z Shi, "Investigation of graphene assisted tunable terahertz metamaterials absorber," *Opt. Mater. Express*, vol. 6, no. 2, pp.331–342, 2016.

[21] Ling X Y, Z Y Xiao, X X Zheng, "Tunable terahertz metamaterial absorber and the sensing application", *Sci. Mater. Electron*, vol. 29, no.1, pp.1–7, 2017.

[22] Fangrong Hu, Yixian Qian, Zhi Li, Junhao Niu, Kun Nie, Xianming Xiong, Wentao Zhang and Zhiyong Peng. "Design of a tunable terahertz narrowband metamaterial absorber based on an electrostatically actuated MEMS cantilever and split ring resonator array", *J. Optic.*, vol.15, no. 5, pp. 055–101, 2013.

[23] Xu W, L Xie, Y Ying, "Mechanisms and applications of terahertz metamaterial sensing: a review," *Nanoscale*, vol. 9, pp.13864–7, 2017.

[24] Salim A, S Lim, "Review of recent metamaterial microfluidic sensors", *Sensors*, vol.18, pp. 232,2018.

[25] Yoo S J, Q H Park, "Metamaterials and chiral sensing: a review of fundamentals and applications." *Nanophotonics*, vol. 8, pp. 249–61,2019.

[26] BX, X Zhai, G Z Wang, WQ Huang, L L Wang, "Frequency tunable metamaterial absorber at deep-subwavelength scale", *Opt. Mater. Express*, vol. 5, pp. 227–235, 2015.

[27] Oszwalldowki M and M. Zimpel, "Temperature dependence of intrinsic carrier concentration and density of states effective mass of heavy holes in InSb," *Journal of Physics and Chemistry of Solids*, vol. 49, pp. 1179–1185, 1988.

[28] Li W, D. Kuang, F. Fan, S. Chang, and L. Lin, "Subwavelength B-shaped metallic hole array terahertz filter with InSb bar as thermally tunable structure," *Applied Optics*, vol. 51, no. 21, pp. 7098–7102, 2012.

Chapter 8

Terahertz Metamaterial Absorbers and Their Applications in Refractive Index Sensing

Pouria Zamzam, Pejman Rezaei*, Amirhossein Norouzi Razani and Seyed Amin Khatami

Electrical and Computer Engineering Faculty, Semnan University, Semnan, Iran

Abstract

Terahertz technology is developing rapidly with new tools and devices. One of these devices is the electromagnetic absorber. The work of these absorbers is based on the absorption of electromagnetic radiation by preventing radiation and reflection of electromagnetic waves. The researchers paid attention to metamaterials because these materials have special and different properties from natural materials such as negative refractive index. Electromagnetic absorbers are divided into narrow-band and wide-band absorbers, depending on the absorption at a specific frequency or in a frequency range, respectively. In designing and manufacturing electromagnetic absorbers, application requirements must be considered first, but in general, features such as the insensitivity of the structure to the incident wave's polarization and incidence angle, make the structure more flexible for different applications. Applications of terahertz absorbers include spectroscopy, imaging, sensing, filtering, etc. Among these applications, absorption sensors have become increasingly important in recent years, as these sensors, which are usually based on the refractive index of different materials, can be used in biomedical and biochemical sensors. In refractive index sensors, the absorption spectra

* Corresponding Author's Email: prezaei@semnan.ac.ir.

In: Metamaterials for Microwave and Terahertz Applications
Editors: B. Appasani, O. Prakash Acharya, A. Vidyakant Jha et al.
ISBN: 979-8-88697-330-3

shifts with the change in refractive index of the surrrounding medium, which can be used to detect the medium. This chapter discusses these absorbers and their application in refractive index sensing. It also discusses the theory behind the design of graphene based absorbers and their application in sensing.

Keywords: perfect metamaterial absorber, terahertz sensor, graphene, metamaterials

1. Introduction

Increasing bit rate and consequently bandwidth increment, as the demand of today's telecommunications, motivates research towards higher frequencies, terahertz (THz), infrared, and optics. Redesigning of well-known communication devices in the microwave frequency range, such as antennas [1-4], absorbers [5-11], filters [12, 13], and sensors [14-22], is a requirement of this process. Research on the subject of electromagnetic absorbers and the study of their applications requires the science of wave propagation and the laws governing the electromagnetic wave (EM Wave). The following are some of the essential points in the science of wave propagation for the search for electromagnetic absorbers: (a) Transmission line theory, which contributes to the basic understanding of the EM wave propagation phenomenon (b) Analytical methods for learning reflection and transmission in EM wave propagation phenomena (c) Different EM wave behaviors (d) EM wave analysis theory using computer simulators (e) Basic knowledge of the materials that make up EM waves (f) Constant measurement of materials (g) Analysis of wave propagation in different environments (h) Concepts of synthetic raw materials and metamaterials, etc. EM waves in different frequency bands start from the radio and microwave spectrums and end at THz frequencies, gamma-ray spectra, and high-energy gamma-ray spectra, which have frequencies up to quadrillion.

Any material that is not available in nature can be called metamaterial. Although this is a simple definition of metamaterials, other definitions are discussed on various topics and sciences. Several examples of these definitions are provided in this chapter. Metamaterials are composite and artificial materials that have been engineered. They can also be described as abnormal, unique, and weird materials. Another definition of metamaterials is that they are usually composed of periodic metal structures whose periodicity is much shorter than their wavelength. Therefore, study and research in metamaterials are interdisciplinary and are used in electromagnetics, optics, optoelectronics, semiconductor engineering, and solid-state physics. The first attempts to find synthetic materials were made in 1898 by Jagadish Chandra Bose, who researched materials with chiral properties. Nevertheless, one of the most important researches in this field dates back to 1968, when Veselago presented a theoretical work on a material with a negative refractive index (NRI) with negative permittivity and permeability [23]. Veslago's work did not receive much attention until 1996, when Pendry succeeded in using an array of metal wires to fabricate a material with a negative permeability coefficient.

They are myriad applications of metamaterials, such as negative refractive index, reverse doppler effects, invisibility cloak, perfect lens, negative refraction, artificial magnetism, transparency, perfect imaging, hyperbolic dispersion, sensing, perfect absorption, antennas, super lensing, switches, modulators, polarization rotation, filters, energy harvesting, multiplexers, etc. These applications can be used over various frequency spectrums, from microwave to optical.

THz waves are part of electromagnetic waves in the frequency range of 0.3 to 3 THz. In terms of wavelength, the range of these waves is between 0.1 mm to 1 mm. It is important to note that THz is a frequency range of electromagnetic waves that have neither the properties of a microwave nor the properties of high-energy waves. Practically THz waves cannot be accessed directly and must be measured by periodic properties of wavelength or energy. The penetration depth of THz radiation is less than microwave and greater than infrared, so THz waves are placed between these two spectra. These waves pass through a wide range of materials such as plastic, wood, cardboard, plaster, and building materials, but these waves are not helpful for telecommunications and are absorbed by water vapor in the Earth's atmosphere. In another category, electromagnetic waves between 0.1 and 10 THz are THz waves. This spectral gap is also known as the THz gap. The main reason for this naming is the lack of an electronic device for measuring the frequency of THz radiation, which cannot be measured digitally via electronic counters that are common today. In addition, the THz frequency range is sometimes referred to as the sub-millimeter band. Interestingly, the study of THz waves has become much more common today, and in the field of THz wave generation, ultra-sensitive detection and manipulation are still in their infancy.

Of course, one of the advantages of THz waves, which has attracted the attention of many researchers and scientists in the last decade, especially in recent years, is the ability to destroy nearly zero of these waves to the living cell due to the non-ionizing nature of THz waves. Another advantage that can be mentioned is that today, due to the use of most applications at frequencies below THz, they are associated with the great challenge of dimensions, bulky equipment, heavy, portable, and expensive. Moving to THz solves some of these problems. THz will revolutionize the transportation industry, allowing cars to be controlled automatically. It will also revolutionize medical imaging, imaging cancerous tissue or images of the internal structure of a tooth, agriculture, and industry.

Absorbers are the most widely used structures to reduce the cross-section of the radar. These absorbers work by absorbing the beams of the incoming EM waves. With the advancement of technology and the ability to design nanoscale structures, there is an effort to make absorbers much thinner and lighter. A suitable method to design such absorbers is the use of metamaterials, which have become popular due to their unique features, such as low weight, volume and cost, and simple structure. It is also possible to design them at THz frequencies. Metamaterial absorbers have been proposed with single-band, multi-band, and broadband absorption characteristics.

Despite the thin structure and unity absorption offered by these structures, one of the disadvantages is their narrow functional bandwidth. The bandwidth of metamaterial absorbers can be increased using multi-layer, multi-resonance, and monolayer structures. In multi-layer structures, many layers of dielectric material are used to achieve broadband absorption. The multiple layers increase the absorber's thickness, making them challenging to use for different applications. Multi-resonant structures are composed of several resonators that give wide absorption bandwidth, although their size is large. As a result, the resonators are combined to form a compact resonator in one layer, which is easy to design and control its resonant frequency. Single-layer structures, with proper optimization, will increase absorption and bandwidth. Although these absorbers show absorption with optimized bandwidth with a simpler geometric structure, the unit cell dimensions are too large for their implementation in conventional applications. These metamaterial absorber structures are generally composed of arrays of unit cells at the top of the substrate and a full metallic plate at the bottom of the substrate.

In this chapter, an attempt is made to examine the performance of these structures at THz frequency. This chapter also examines THz sensors, one of the applications of THz absorbers.

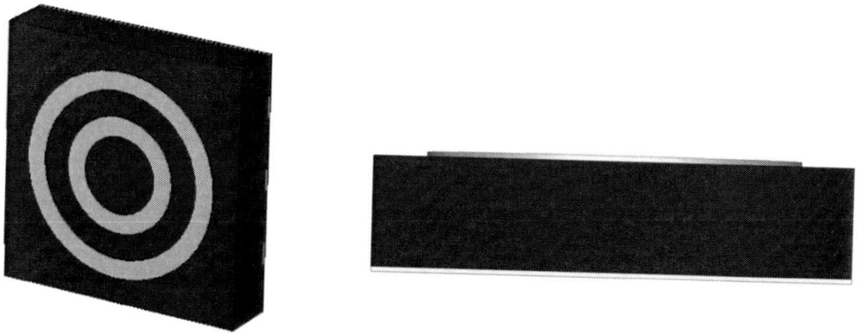

Figure 1. Perspective and side view of a classic absorber consisting of three main layers.

2. An Overview of THz Metamaterial Absorber

A metamaterial absorber generally consists of three layers: a metallic layer, a dielectric layer, and a metallic layer. The goal is to achieve an absorption peak above 80% in the THz spectrum, absorbing a good and acceptable percentage of electromagnetic waves in this frequency range. In addition, the absorption should be insensitive to polarization and incident angle. The competition for a high percentage of absorption has made some structures unique and attractive, and now there are structures that, in addition to not being sensitive to the polarization of incident waves, have several absorption bands, each above 99%. In general, such structures are called perfect THz absorbers, insensitive to polarization, and have a wide incident angle. The meaning of a perfect or nearly perfect absorber is that it has an absorption peak above 98%.

2.1. Classical Absorber Structures

A classical THz absorber structure consisting of three layers is shown in Figure 1 [36]. The upper and lower layers are made of a perfect conductor. The middle layer of this structure is made of polyamide metamaterial with a refractive index of n. Note that the following formula is used to obtain the dielectric constant from the refractive index equation.

$$n = n_r(\omega) + ik(\omega) \tag{1}$$

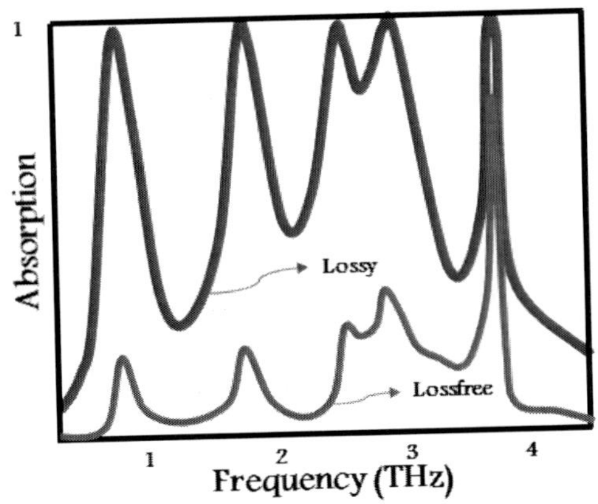

Figure 2. Comparison of absorption curves in the lossy structure and loss-free one.

$$\varepsilon_r = n_r^2 - k^2 \tag{2}$$

$$\varepsilon_{Im} = 2n_r k \tag{3}$$

According to the above formulas, the relative permeability coefficient of the environment and the loss tangent $\left(\tan(\delta) = \varepsilon_{Im}/\varepsilon_r\right)$ can be calculated.

The middle dielectric layer is used as a loss layer for absorption. This can be understood from Figure 2, which clearly that the lossy nature of the dielectric is responsible for the wave absorption [36].

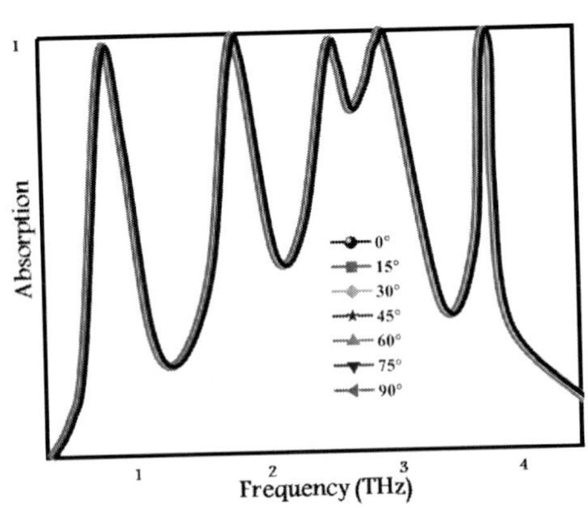

Figure 3. Dependence of the absorption spectra on the polarization angles of the normally incident wave.

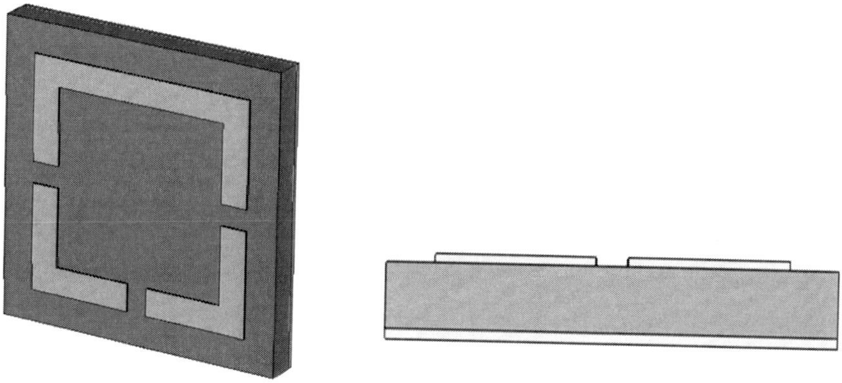

Figure 4. Perspective view and side view of a classic THz absorber structure.

It should be noted that the structure's lossy dielectric or non-lossy depends on whether the imaginary part of the middle layer is zero. In addition to having five absorption bands, this structure is polarization insensitive, as shown in Figure 3. This is because of the symmetrical nature of the design. Thus, for the design to be insensitive to polarization angle, it should be symmetrical.

Another three-layered classical absorber structure with four absorption bands is shown in Figure 4 [37]. The middle layer is sandwiched between two perfect conductors. The middle layer is made of polyamide and is lossy. The absorption spectrum can be seen in Figure 5.

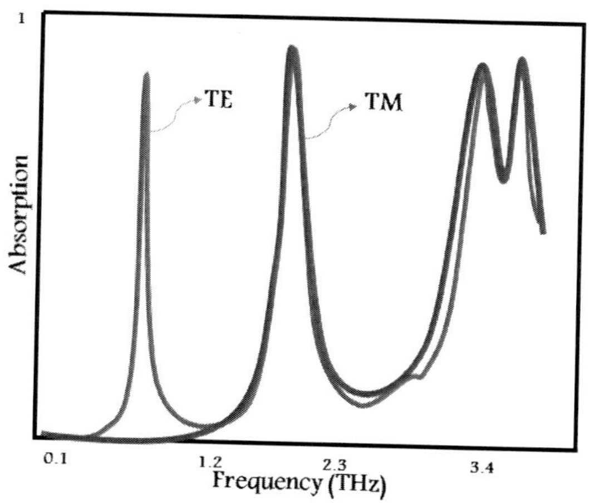

Figure 5. Absorption spectra in both TE and TM modes.

Figure 6. Perspective view of proposed classic THz absorber structure.

In [38], the structure is made of three layers of metal-dielectric-metal, the middle layer of which is dielectric, sandwiched between two perfect electric conductors. Figure 6 clearly shows all the points. The absorption curve of this structure can be seen in Figure 7. This structure has four absorption bands, measured once for the lossless state of the middle layer and once with the loss of the middle layer.

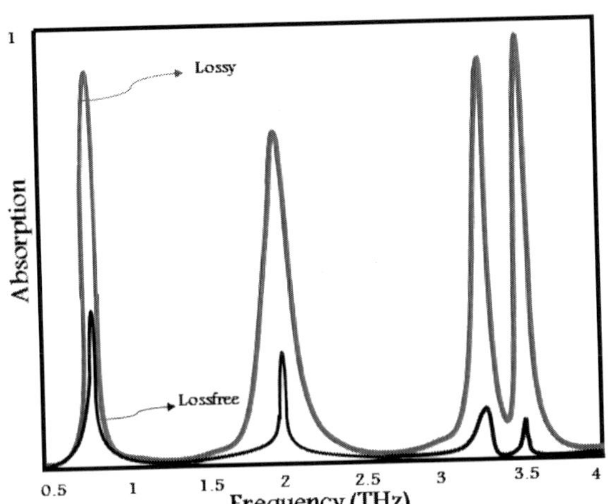

Figure 7. Comparison of absorption curves with loss and no loss.

Figure 8. The schematics of a single unit cell.

As an excellent conclusion to the study of classical absorber structures, a new classical metamaterial perfect absorber unit cell having three layers will be discussed [39]. It is a three-layered metal-insulator-metal design that has a dielectric of varying height. The structure is shown in Figure 8.

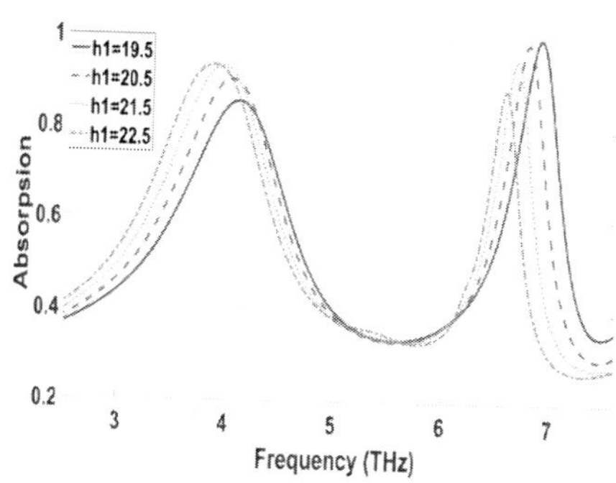

(a)

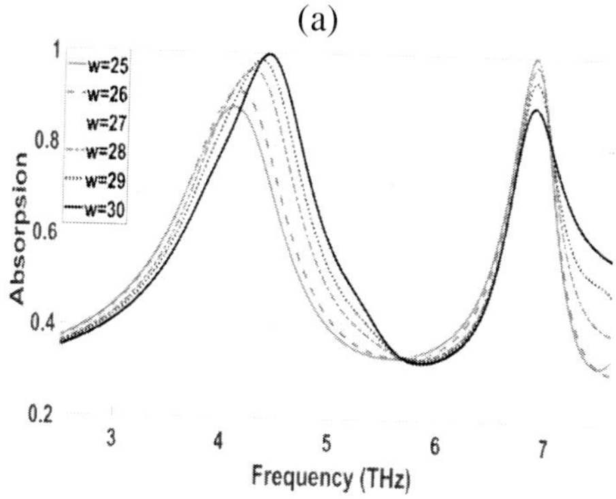

(b)

Figure 9. Absorption spectra of the presented MPA (a) With different column height, (b) With different column width.

This structure has two perfect absorption bands, which change with the dielectric height and column width as shown in Figure 9.

The equivalent circuit shown in Figure 10. The basic formula of the absorption equation is:

$$A(\omega)=1-T(\omega)-R(\omega) \tag{4}$$

where $A(\omega)$ is the absorption coefficient, $T(\omega)$ is the transmission coefficient, and $R(\omega)$ is the reflection coefficient. As Figure 10 shows the underlying layer of the structure is a continuous conductor and is thicker than the skin depth, therefore, no THz wave passes through the structure, and practically $T(\omega)$ is considered zero. So, part *A* in the circuit is equivalent to the display of the same conductive substrate, considered a short circuit, with all the power reflected. Part *B* shows the transmission line with peripheral impedance Z_d, the same as a dielectric with height *m*. Therefore, changes in height *m* in the equilibrium circuit are effective, but part *C* consists of Z_m and Z_g, as two parallel impedances. Z_m is called the dielectric-metal interference impedance. Z_g is the impedance of the conducting layer on top of the dielectric, which always appears as *RLC* circuit. Part *D* consists of two parallel impedances of the dielectric columns and the impedance of air or free space between the two dielectrics, which depends on the height $h1$. Part *E* consists of two *RLC*s parallel to the top layer. Finally, part *F* shows the free space impedance or Z_0. The absorption of the proposed absorber can be obtained from the following equations.

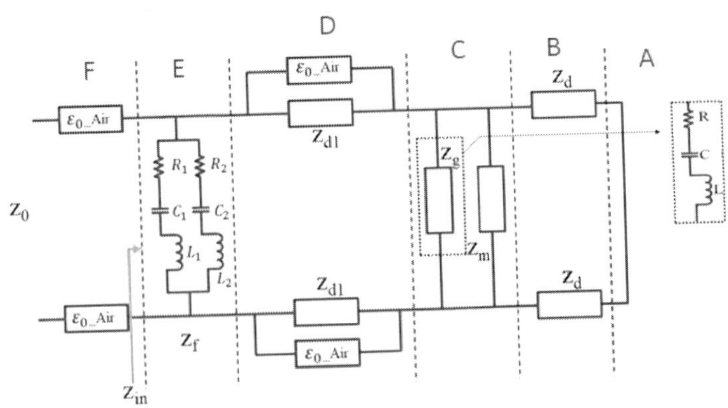

Figure 10. Equivalent circuit model of proposed PMA.

$$A(\omega) = 1 - \left[\left| \frac{z_{in}(\omega) - z_0}{z_{in}(\omega) + z_0} \right| \right]^2 \tag{5}$$

z_{in} is the input PMA impedance, expressed as:

$$z_{in}(\omega) = \frac{1}{z_f(\omega)} + \frac{1}{z_{d1}(\omega) \| z_0} + \frac{1}{z_g(\omega)} + \frac{1}{z_m(\omega)} + \frac{1}{z_d(\omega)} \tag{6}$$

Here in:

$$z_f(\omega) = \frac{1}{R_1 + j\omega L_1 + \frac{1}{j\omega C_1}} + \frac{1}{R_2 + j\omega L_2 + \frac{1}{j\omega C_2}} \tag{7}$$

$$z_{d1}(\omega) = j\sqrt{\frac{\omega_r \omega_0}{\varepsilon_r \varepsilon_0}} \tan(kd_1) \tag{8}$$

$$z_g(\omega) = \frac{1}{\sigma_g} \tag{9}$$

$$z_m(\omega) = \frac{z_{d1}(\omega)}{\frac{2\sqrt{n_1 n_2}}{T} - (n_1 + n_2)} \tag{10}$$

$$z_d(\omega) = j\sqrt{\frac{\omega_r \omega_0}{\varepsilon_r \varepsilon_0}} \tan(kd) \tag{11}$$

$$k = \frac{k_0}{\sqrt{\varepsilon_r k_r}} \tag{12}$$

where k is the wavenumber, ω_r is permeability and ε_r is the real dielectric constant of the proposed spacer, k_0 is the wavenumber of the free space, ω_r is the permeability of the free space, and ε_r is the free space relative permittivity. Also, Z_0 is air impedance (about 377 ohms), and n_1 and n_2 are the dielectric and gold material refractive indexes.

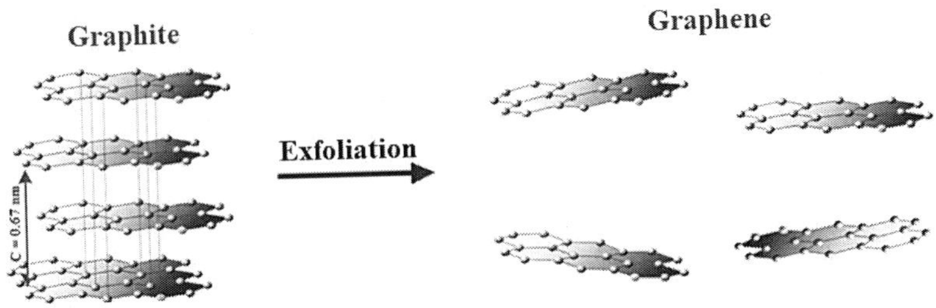

Figure 11. Creating graphene by breaking the bond between the layers of carbon atoms.

2.2. Graphene-Based Absorbers

Graphene is a metamaterial made from graphite. As shown in Figure 11, graphite consists of layers of carbon atoms. Each layer of carbon atoms is bonded with a strong bond to the adjacent layer, and by breaking these bonds (called exfoliation), graphene is created. Graphene is a single layer of carbon atoms and is therefore called a two-dimensional material. It was first developed in 2004 by Andre Gaim and Konstantin Novoself at the University of Manchester. The two scientists received the Nobel Prize in Physics in 2010 for building this two-dimensional metamaterial.

2.2.1. Applications and Unique Properties of Graphene

The unique properties of graphene led to its application in electrical engineering, photonic engineering, materials engineering, and applied medicine. Among its applications in the field of electrical engineering in the field of field and wave communications, the most important of which are sensors, attractions, antennas, and optics. Due to graphene's conductivity, this metamaterial can be used to make electromagnetic shields that can absorb different frequencies of waves. Also, graphene finds its use as the antennas operating in the THz field are small in size (nanometer range). In addition, graphene, due to its adjustability, allows us to make antennas with controllable polarization. Graphene is also of interest to researchers in manufacturing optical instruments, including optical computers and wireless computers. In addition to its small size and two-dimensional properties, graphene's conductivity can be adjusted by external bias voltage and chemical doping, receiving much attention in recent years [24-31]. The unique properties of graphene that distinguish it from other materials are presented in Table 1.

Table 1. Unique properties of graphene

Properties	**Further details**
High conductivity	Graphene is six times more conductive than copper.
Very low weight	One square meter of graphene weighs 7.7 mg.
High strength	Interestingly, graphene, with its low weight, is 100 times stronger than steel.
Transparency	Graphene is a transparent metamaterial that allows us to use it in making screens.

2.2.2. Surface Conductivity of Graphene

The graphene surface conductivity can be calculated using the Kubo formula as defined in the following [32-34]:

$$\sigma_{\text{intra}}\left(\omega, E_f, \tau, T\right) = \sigma_{\text{intra}}\left(\omega, E_f, \tau, T\right) + \sigma_{\text{inter}}\left(\omega, E_f, \tau, T\right) \tag{13}$$

$$\sigma_{\text{intra}}\left(\omega, E_f, \tau, T\right) = \frac{2k_B e^2 T}{\pi h^2} \ln\left[2\cosh\left(\frac{E_f}{2k_B T}\right)\right] \frac{i}{\omega + i\tau^{-1}} \tag{14}$$

$$\sigma_{\text{inter}}\left(\omega, E_f, \tau, T\right) = \frac{e^2}{4h}\left[H\left(\frac{\omega}{2}\right) + \frac{4i\omega}{\pi}\int_0^{\infty} \frac{H(\varepsilon) - H\left(\frac{\omega}{2}\right)}{\omega^2 - 4\varepsilon^2} d\varepsilon\right] \tag{15}$$

where:

$$H(\omega) = \frac{\sinh\left[\frac{\hbar\varepsilon}{k_B T}\right]}{\cosh\left[\frac{E_f}{k_B T}\right] + \cosh\left[\frac{\hbar\varepsilon}{k_B T}\right]}$$

ω Angular frequency
E_f Fermi energy level
τ Relaxation time
T Temperature (K)
k_B Boltzmann's constant
E Electron charge
$\hbar$ reduced Planck's constant

Based on the Pauli Exclusion Principle, in the THz band, $E_f >> k_B T$, and the conductivity of the inter-band becomes insignificant [35]. Therefore, the equation of graphene conductivity due to the Drude-like model is:

$$\sigma\left(\omega, E_f, \tau, T\right) = \frac{e^2 E_f}{\pi h^2} \frac{i}{\omega + \frac{i}{\tau}} \tag{16}$$

The relaxation time of graphene is calculated using the following equation:

$$\tau = \frac{\mu E_f}{e V_f^2} \tag{17}$$

where μ and V_F represents the electron mobility and Fermi velocity, respectively. Furthermore, the effective thickness of graphene is $t_g = 1$ nm, and its dielectric constant can be calculated using the following equation:

$$\varepsilon_r = 1 + \frac{j\sigma(\omega)}{\omega \varepsilon_o t_g} \tag{18}$$

Where ω is the angular frequency and ε_0 is the vacuum permittivity.

2.2.3. Graphene Bias

As mentioned, one of the properties of graphene is its adjustable conductivity. There are various methods for regulating the conductivity of graphene, the most common of which is adjusting the conductivity of graphene by changing the Fermi level of graphene, which can be done by two methods of chemical doping and electrical gating.

Figure 12 shows the electrical gating method for the structure of electromagnetic absorbers, which is performed using an external DC-bias voltage and an electrode layer, which is usually made of polysilicon. It should be noted that the thickness of the electrode layer should be considered as small as possible so as not to interfere with the performance of electromagnetic absorbers.

According to Figure 12, the chemical potential of graphene (E_F) can be adjusted according to equation (19) using VG voltage.

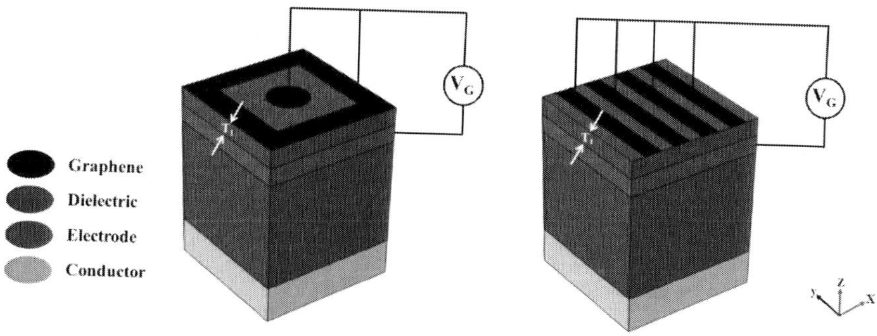

Figure 12. External DC-bias voltage of graphene.

$$V_G = \frac{n_s e T_1}{\varepsilon_o \varepsilon_r} \tag{19}$$

where T_1 is the thickness of the dielectric layer between the graphene layer and the electrode layer, ε_r is the relative permittivity of dielectric, and n_s can be calculated by the following equation:

$$n_s = \frac{2}{\pi h^2 V_f^2} \int_0^\infty \varepsilon \left[f_d\left(\varepsilon, \mu_c, T\right) - f_d\left(\varepsilon + 2\mu_c, \mu_c, T\right) \right] d\varepsilon \tag{20}$$

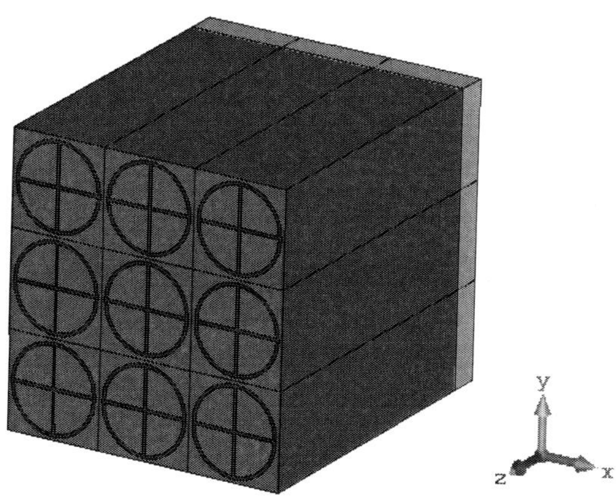

Figure 13. Perspective view of 3 × 3 array graphene THz absorber structure.

where V_F = 9.5 × 10^5 m/s is the Fermi velocity, ε represents energy, $f_d\left(\varepsilon,\mu_c,T\right)$, and $f_d\left(\varepsilon+2\mu_c,\mu_c,T\right)$ Equations (21) and (22) can be determined, respectively.

$$f_d\left(\varepsilon,\mu_c,T\right)=\frac{1}{e^{\frac{(\varepsilon-\mu_c)}{k_BT}}+1} \tag{21}$$

$$f_d\left(\varepsilon+2\mu_c,\mu_c,T\right)=\frac{1}{e^{\frac{(\varepsilon+\mu_c)}{k_BT}}+1} \tag{23}$$

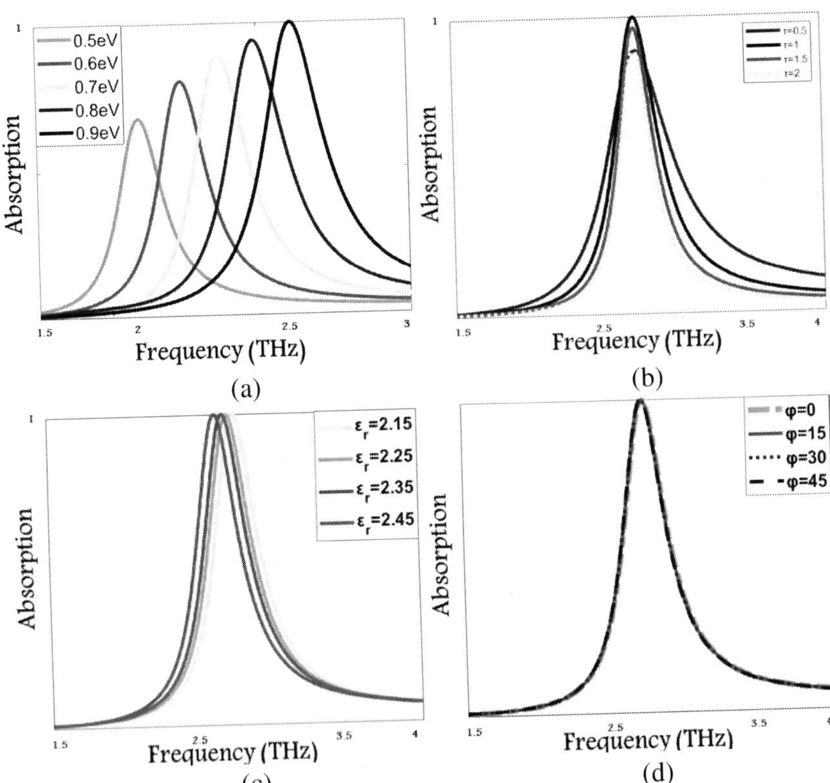

Figure 14. Absorption spectra of the presented MPA (a) with various Fermi level, (b) with different relaxation times (τ) varies from 0 to 2 ps, (c) with different dielectric material properties, (d) with different polarization angles of the normally incident wave.

2.2.4. Metamaterial Absorber with Graphene

A graphene absorber is proposed in [40], the upper layer of which is made of graphene, the lower layer of which is made of aluminum, and the middle layer is made of SiO_2, and is shown in Figure 13.

This explains the use of graphene as the top layer. From the absorption spectra of this structure in Figure 14 (a-d), this structure has achieved a higher absorption percentage by varying the properties of graphene.

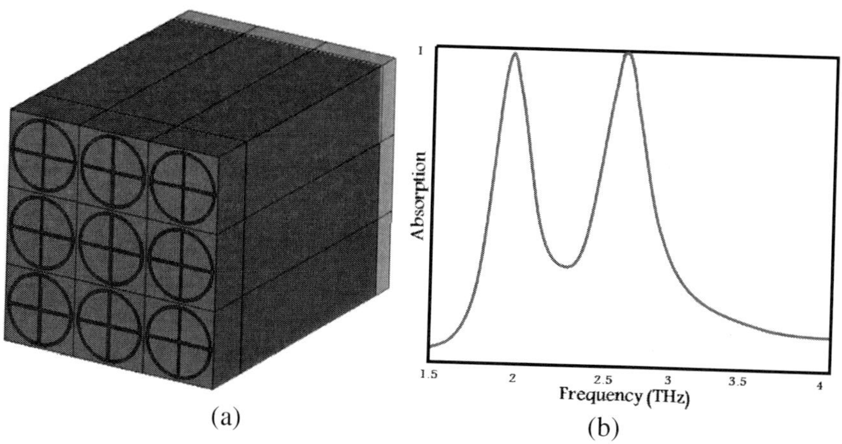

Figure 15. (a) The perspective of the proposed dual-frequency tunable absorber, which is composed of double-layer graphene metasurface. (b) The absorbance spectra of the proposed absorber.

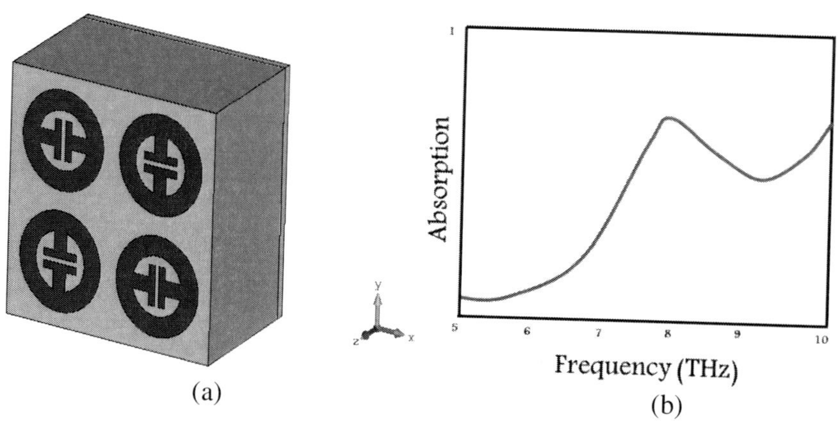

Figure 16. Perspective of the proposed single-layer absorber, (b) The absorption spectra of the proposed absorber.

When a multi-layer structure is used, double band the absorption is achieved. The schematic of the proposed dual-frequency tunable absorber and the absorbance spectra of the proposed absorber is shown in Figure 15 (a, b).

In [41], the multi-layer technique is used to improve the absorption performance without symmetry in the structure. Figure 16 shows the perspective view of the single-layer structure with the absorption curve, and Figure 17 shows the perspective view of the double-layer and triple-layer structure and the improvement of the absorption performance.

In [42], one of the essential and necessary conditions for using an additional layer to obtain more absorption bands in metamaterial THz absorbers is determined. If the form or pattern of the upper layer has cross-symmetry, a quad-band absorber will be achieved by adding another layer. First, a perfect dual-band graphene absorber structure is introduced, which is shown in Figure 18.

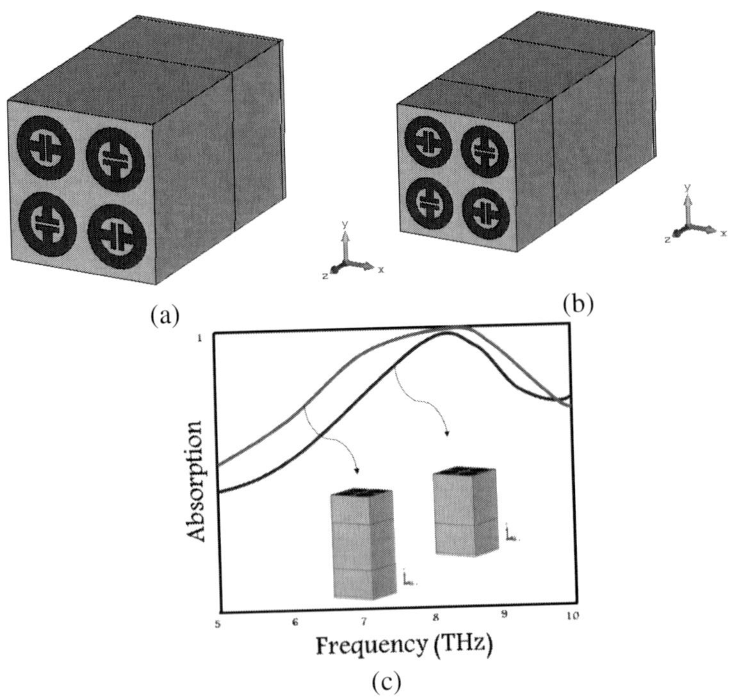

Figure 17. (a) Perspective of the proposed double-layer absorber, (b) Schematic of the proposed triple-layer absorber, (c) The absorption spectra of the proposed absorber.

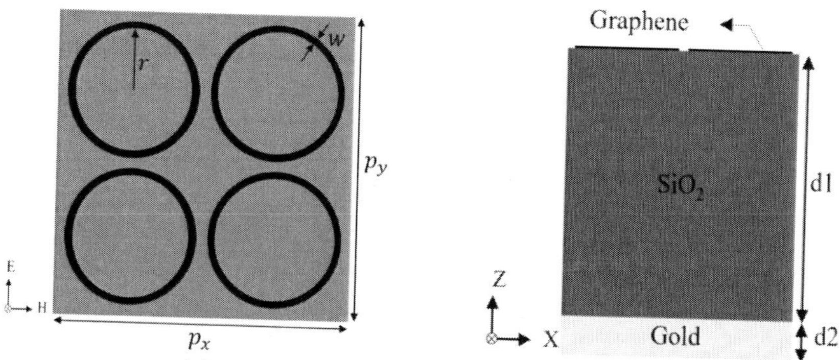

(a) (b)

Figure 18. (a) Top view of model 1. (b) Side view. The geometric parameters of the designed structure are: d1=10 μm, d2=100 nm, $p_x = p_y = 3\ \mu m$, $r = 650\ nm$, $w = 50\ nm$.

The absorber's bottom layer is gold with a conductivity of $\sigma_{gold} = 4.56 \times 10^7$. The layer thickness (d2) is 100 nm. As shown in Figure 18, the middle layer is made of SiO_2 with a permittivity of 2.25 and a thickness (d1) of 10 μm. The top layer is made up of a very thin graphene arrangement, with r1 and r2, inner and outer ring shape radius, respectively. The periodicity of the proposed structure is $p_x = p_y = 3\ \mu m$. Four different models with different top layer arrangements were designed as shown in Figure 19, and their corresponding absorption spectra are shown in Figure 20.

(a) (b) (c)

Figure 19. (a) Top view model 2, (b) Model 3, (c) Model 4, (d) Model 4. The geometric parameters of the structure are: d1 = 10 μeqm, d2 = 100 nm, $p_x = p_y = 3\ \mu m$, $r = 650\ nm$, $w = 50\ nm$, L = 100 nm.

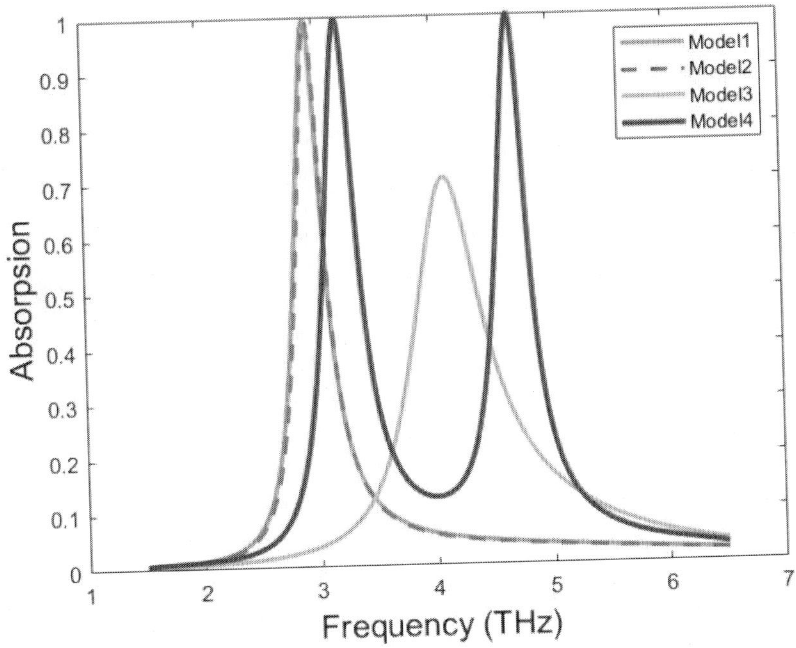

Figure 20. The absorbance spectra of the proposed absorber for different Models.

As can be seen in Figure 20, the absorption percentages of models 1 and 2 are very similar with very little difference. In model 1, the absorption peak at 2.93 THz is 99.5%. Also, the highest absorption of model 2 at 2.94 THz is 99.38%. Model 3 offers an absorption percentage of 70.46% at a frequency of 4.1 THz, but model 4 is the best. This model is a perfect dual-band absorber structure with 99.61% and 99.95% absorption percentages at 3.2 and 4.7 THz, respectively. The important point is that the structure's shape has a cross-symmetry, so adding another layer of dielectric with a thickness of d3 = 2 μm and surface graphene (another layer) will result in a quad-band structure, as shown in Figure 21. This structure is the same as the previous structure and is made of the same dielectric and graphene materials and has four absorption bands of 99.88%, 98.28%, 99.58%, and 99.99% at frequencies of 2.7, 3.2, 3.99, and 4.96 THz, respectively. Figure 22 shows the absorption spectra of this model.

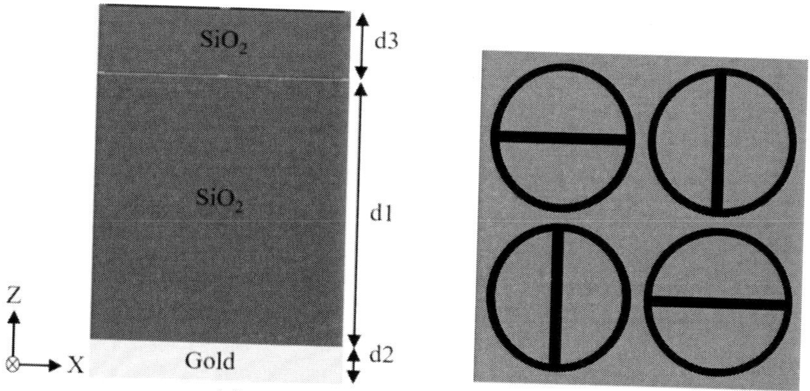

Figure 21. Schematic of the proposed quad-frequency tunable absorber, composed of double-layer graphene metasurface. (a) side view (b) top view.

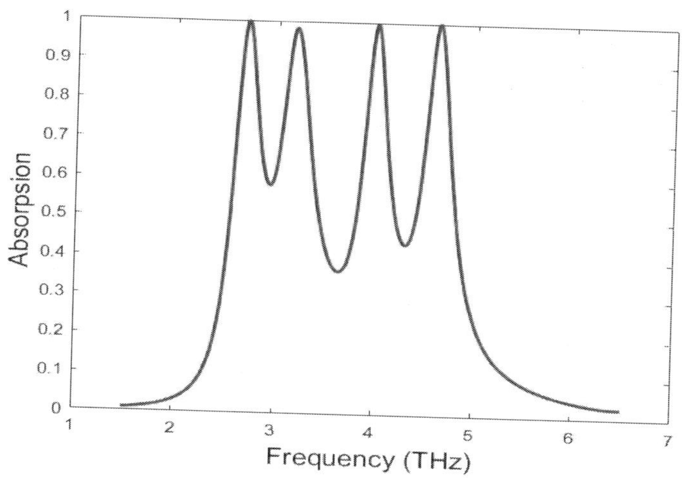

Figure 22. The absorbance spectra of the proposed absorber (Fermi level = 0.9 eV, $\tau = 1$ ps).

Another property of this structure, in addition to being a perfect quad-band, is insensitivity to the polarization angle of the incident electromagnetic wave. As shown in Figure 23(a), the absorption curves change negligible concerning changes in polarization angle of the arrived wave. In addition, this structure has a wide incident angle, which is shown in Figure 23(b), and a property in this structure.

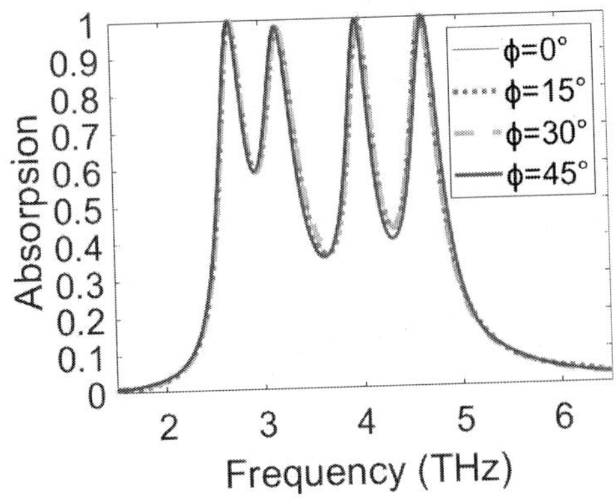

(a)

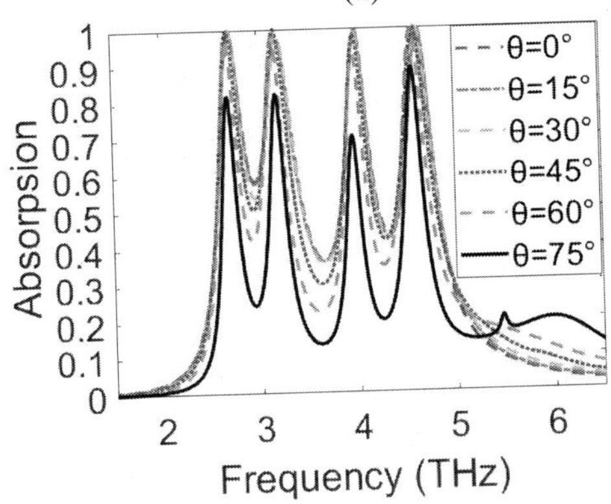

(b)

Figure 23. (a) Dependence of the absorption spectra on the polarization angles of the normally incident wave. (b) Absorption spectra under different incidence angles with TE radiation.

A comparison of the performances of some multi-band absorbers reported in recent years is presented in Table 2.

3. Absorption-Based Refractive Index Sensors

THz absorbers can be used to design sensors. Sensors are devices that respond to a specific physical stimulus and produce a measurable electrical signal. Sensors can be electrical, magnetic, optical, or mechanical. Recent advances in micro-tools, microstructures, digital signal processing, and chemical processes have developed low-cost micro-sensors. Micro-sensors have many applications in medicine, aerospace, military equipment, telecommunications, etc.

Good sensors are only sensitive to the property of measurement and are not sensitive to the rest of the properties in use and do not affect the property they are measuring. Sensors have different operations. Sensors can operate based on refractive index (RI) changes, changes in electrical resistance, changes in capacitance, electrical induction, thermoelectric effect, Hall effect, photoelectric effect, piezoelectric effect, etc.

Table 2. Comparison of the performances of some recent multi-band absorbers

References	Frequency (THz)	Peak Numbers	Average Absorption	Polarization Insensitive	Incident Angle (deg)
[24]	2.7, 3.2, 3.99, 4.6	4	99.43%	Yes	0-60
[43]	0.114, 0.181	2	98.95%	Yes	0-45
[44]	0.85, 3.37	2	95.15%	Yes	0-80
[45]	0.77,1.13, 1.53, 2.06	4	97.89%	Yes	0-90
[46]	0.53, 0.84, 0.98	3	92.73%	Yes	0-45
[47]	147, 188, 269	3	98.53%	No	No
[48]	0.51, 1.25, 2.1, 2.34	4	98.87%	-	-
[49]	1.6, 4.42, 4.94	3	99.06%	Yes	0-60
[50]	0.41, 0.75	2	94.8%	Yes	No
[51]	7.75, 14.85, 23.43	3	99%	Yes	-
[52]	0.94, 1.56	2	99.23%	Yes	0-50
[53]	469, 497.5	2	99.25%	Yes	0-75
[54]	1.59, 2.45, 3.05, 4.41, 5.11	5	97.35%	-	-
[55]	22.33, 27.59	2	98.77%	Yes	0-90

3.1. Applications of Absorption-Based RI Sensors

The absorption-based sensors work based on the refractive index changes of the measured sample [56]. Recently, a lot of research and studies have been done on absorption-based sensors, and many advances have been made in this field. Two very important applications of absorption-based sensors are biomedical and biochemical sensing.

Many diseases, such as cancer, have no definitive cure, and only early detection can help treat cancer patients. In addition, some diseases, such as malaria, do not have specific symptoms; therefore, the error rate of clinical diagnoses of such diseases is high, and with the slightest mistake, the patient's life may be endangered. That is why the need for a biomedical sensor is always felt.

Biomolecules will be damaged by high-energy photons and significant loss in the infrared, visible, and ultraviolet regions, so optical sensors are not a good choice for biomedical sensing. Therefore, RI sensors based on THz metamaterial perfect absorbers (MPA) are good candidates for biomedical sensing.

3.2. Evaluation of Sensors

Three basic parameters, sensitivity (S), the figure of merits (FoM), and Q-factor, are usually calculated and examined to evaluate the performance of absorption-based RI sensors. Sensitivity reflects the response level of the sensor to the external factors. Q-factor is defined to describe the resonator's energy-storing or wavelength selecting ability, which is an essential property. The FOM is the sensitivity (S) ratio and the absorption curve's full width half maximum (FWHM) ratio. It is an inclusive parameter to assess the performance of an absorption-based RI sensor. Sensitivity, FOM, and Q-factor can be calculated by equations 32, 33, and 34, respectively.

$$S = \frac{\Delta f}{\Delta n} \tag{24}$$

$$FoM = \frac{S}{FWHM} \tag{25}$$

$$Q_{factor} = \frac{f}{FWHM} \tag{26}$$

where Δf, Δn, and f are the frequency shift of the absorption peak, the RI change, and the absorption peak frequency, respectively. In addition, the refractive index is related to the relative permittivity by $n^2 = \varepsilon_r$. According to Equation (25), sensitivity (S) has a linear relation with FoM, i.e., to increase FoM, the sensitivity (S) of the sensor must be increased.

3.3. A Graphene-Based Refractive Index Sensor

A THz metamaterial absorber is discussed that can be used as a refractive index sensor is shown in Figure 24. The top layer is made up of graphene, and the bottom layer is made of gold. The spacer layer between gold and graphene is made of SiO_2The top graphene layer is designed as an array of patches. The test medium (analyte) should be added to the absorber structure, as shown in Figure 24 (b). When the refractive index of the test medium changes, its absorption spectra changes. The absorption spectra of the proposed sensor is shown in Figure 25 for different refractive indexes of the analyte.

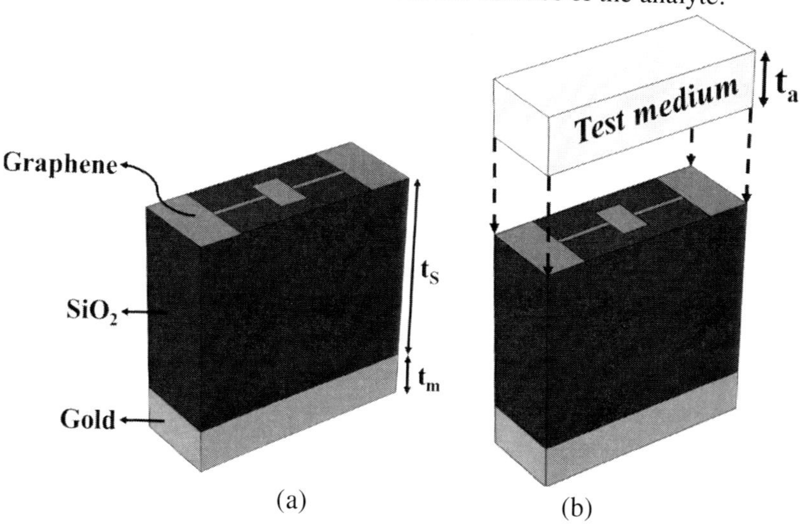

Figure 24. (a) The schematic view of the reported absorber. (b) Adding the test medium to the reported THz sensor based on the perfect metamaterial absorber.

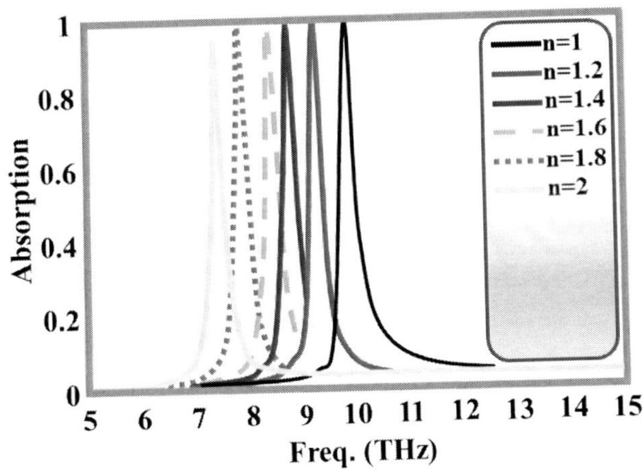

Figure 25. The absorption spectra of the reported sensor for different refractive indices of the test medium with $t_a = 0.8$ µm.

Table 3. The sensing parameters of the reported sensor for different values of the analyte thickness (t_a)

t_a (µm)	f (THz)	S (THz/RIU)	FWHM (THz)	FoM (RIU^{-1})	Q
0.2	8.95	1.51	0.27	5.59	33.15
0.4	9.04	2.52	0.275	9.18	32.54
0.6	9.05	5.05	0.278	18.17	31.94
0.8	9.09	6.56	0.265	24.78	33.32
1	9.22	3.03	0.252	12.02	36.58
1.2	9.25	2.27	0.305	7.45	30.13

From Figure 25, it can be seen that the resonance peaks shift with the change in refractive index. Table 3 presents the sensitivity and resonance frequency of the reported sensor for different analyte thickness (t_a) values, as can be seen from this table, when thickness $t_a = 0.8$ µm, the reported sensor has the highest sensitivity. Therefore, the choice of analyte thickness should be optimal.

As mentioned, to increase FoM, the sensor's sensitivity has to be increased. In order to increase the sensitivity of the reported sensor, the number of graphene patches has been increased, as shown in Figure 26. According to Table 4, with the increase in the number of graphene patches, the sensitivity of the reported sensor has increased from 6.56 THz/RIU to 22.72 THz/RIU, and following this increase, as expected, the FoM has increased from 24.78 RIU^{-1} to 34.96 RIU^{-1}. However, it should be noted that the Q-factor in the sensor reported with three graphene patches has decreased from 33.32 to 14.32 compared to the sensor reported with one graphene patch, and the reason for this decrease is the increase of FWHM, which according to Equation 34 reduces the Q-factor.

3.4. Applications of Reported Sensor

As mentioned, biomedical sensing is one of the applications of absorption-based THz sensors. In this application, first, the refractive index of the measurement sample must be determined according to different methods. For example, if a sensor is used to detect a specific cancer cell, the RI of the cancer cell's measurement sample must be determined. Usually, well-known diseases have different stages, in each stage of which the RI of the measurement sample will be different, so by determining the sample RI with a minimum error coefficient, the disease stage can be detected with RI absorption-based sensors.

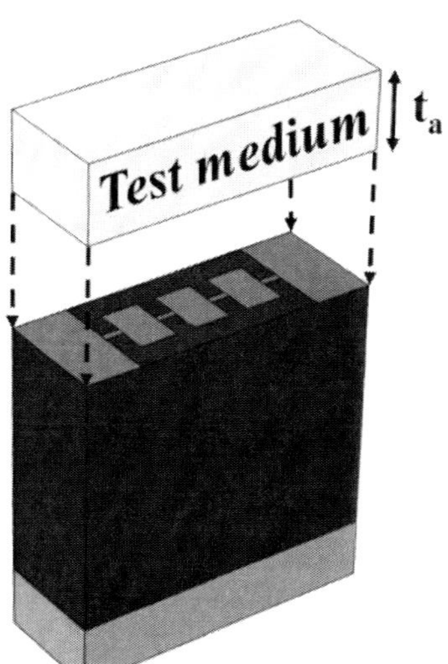

Figure 26. The reported THz sensor with an array of three graphene patches located on top of the structure.

Table 4. The sensing parameters of the reported sensor with three graphene patches

f (THz)	S (THz/RIU)	FWHM (THz)	FoM (RIU^{-1})	Q
9.4	22.72	0.65	34.96	14.32

Table 5. The sensing parameters of the reported sensor for diagnosis of malaria

Malaria Stage	n	S (THz/RIU)	FoM (RIU^{-1})
Schizont	1.373	7.33	27.16
Tropozoite	1.383	10.58	36.51

The parameters of the sensor reported has been examined for the diagnosis of glucose and malaria, and after examination, it was found that this sensor is a good candidate for sensing. Malaria is an infectious disease that affects red blood cells (RBCs). This infectious disease has different stages: the ring stage, trophozoite stage, and schizont stage. The refractive indexes of a healthy RBC, RBC in ring stage, RBC in trophozoite stage, and RBC in schizont stage have already been determined to be 1.399 ± 0.006, 1.395 ± 0.005, 1.383 ± 0.005, 1.373 ± 0.006 respectively.

Table 5 presents the sensitivity and FoM of the reported sensor for RBCs in trophozoite and schizont stages. As can be seen from this table, the reported sensor has a very high sensitivity for detecting different stages of malaria. To conclude, it should be said that absorption-based sensors in the THz frequency range can be used as sensors in various fields such as the food industry, biochemical sensing, biomedical sensing, etc. Since biomolecules will be damaged by high-energy photons and significant losses in the infrared, visible, and ultraviolet regions, THz sensors can be an excellent alternative to optical sensors for biomedical sensing.

Conclusion

This chapter discusses the design of THz metamaterial absorbers and their application in refractive index sensing. The classical design approach used ordinary metals to achieve absorption; lately, graphene is used for EM wave absorption. Graphene-based THz absorber design has been discussed. It has been shown that symmetry checking can lead to improvement in performance. The THz absorber can be used for sensing the refractive index of the surrounding medium. A graphene-based absorber is proposed that can find application in the detection of malaria. The non-ionizing nature of the THz waves can play an important role in the development of future biomedical sensors based on THz metamaterial absorbers.

Acknowledgment

The authors acknowledge the Semnan University for their beneficial and professional help. Also, the authors would like to thank Dr. Bhargav Appasani, Mr. Sina Kiani, Dr. Shiva Khani, Mr. Mohsen Jafari Chashmi, Mrs. Maryam Shayestehnezhad for their constructive comments.

References

[1] Chashmi, Mohsen Jafari, Pejman Rezaei, and Narges Kiani. (2020). "Y-shaped graphene-based antenna with switchable circular polarization." *Optik*, 200, 163321.

[2] Jafari Chashmi, Mohsen, Pejman Rezaei, and Narges Kiani. (2020). "Polarization controlling of multi resonant graphene-based microstrip antenna." *Plasmonics*, 15, no. 2, 417-426.

[3] Kiani, Sina, Pejman Rezaei, and Mina Fakhr. (2021). "A CPW-fed wearable antenna at ISM band for biomedical and WBAN applications." *Wireless Networks*, 27, no. 1, 735-745.

[4] Chashmi, Mohsen Jafari, Pejman Rezaei, and Narges Kiani. (2019). "Reconfigurable graphene-based V-shaped dipole antenna: From quasi-isotropic to directional radiation pattern." *Optik*, 184, 421-427.

[5] Norouzi-Razani, Amirhossein, and Pejman Rezaei. (2022) "Broadband polarization insensitive and tunable terahertz metamaterial perfect absorber based on the graphene disk and square ribbon." *Micro and Nanostructures*, 163, 107153.

[6] Zamzam, P., Rezaei, P., Khatami, A., and Mousavirazi, Z. (2022). Convertible perfect absorber with single ring resonator: Tunable single band infrared/dual band visible, Modeling Simulate. *Elec. Electron. Eng.*

[7] Norouzi-Razani, Amirhossein, and Pejman Rezaei. (2022). "Multiband polarization insensitive and tunable terahertz metamaterial perfect absorber based on the heterogeneous structure of graphene." *Optical and Quantum Electronics*, 54, no. 7, 1-16.

[8] Verma, V. K., Mishra, S. K., Kaushal, K. K., Gupta, N., and Appasani, B. (2020). "An octaband polarization insensitive terahertz metamaterial absorber using orthogonal elliptical ring resonators." *Plasmonics*, 15, no. 1, 75-81.

[9] Mohanty, A., Acharya, O. P., Appasani B., and Mohapatra, S. K. (2018). "A multi-band terahertz metamaterial absorber based on a Π and U-shaped structure." *Photonics and Nanostructures-Fundamentals and Applications*, 32, 74-80.

[10] Pelluri, Rahul, and Bhargav Appasani. (2017). "Genetic algorithm optimized X-band absorber using metamaterials." *Progress In Electromagnetics Research Letters*, 69, 59-64.

[11] Ghods, Mohammad Mahdi, and Pejman Rezaei. (2018). "Ultra-wideband microwave absorber based on uncharged graphene layers." *Journal of Electromagnetic Waves and Applications*, 32, no. 15, 1950-1960.

[12] Khani, Shiva, Mohammad Danaie, and Pejman Rezaei. (2019). "Size reduction of MIM surface plasmon based optical bandpass filters by the introduction of arrays of silver nano-rods." *Physica E: Low-Dimensional Systems and Nanostructures*, 113, 25-34.

[13] Khani, Shiva, Mohammad Danaie, and Pejman Rezaei. (2019). "Tunable single-mode bandpass filter based on metal–insulator–metal plasmonic coupled U-shaped cavities." *IET Optoelectronics*, 13, no. 4, 161-171.

[14] Kiani, S., Rezaei, P., and Fakhr, M. (2021). "An overview of interdigitated microwave resonance sensors for liquid samples permittivity detection." *Interdigital Sensors*, 153-197.

[15] Banerjee, Sagnik, Uddipan Nath, Purba Dutta, Amitkumar Vidyakant Jha, Bhargav Appasani, and Nicu Bizon. (2021). "A theoretical terahertz metamaterial absorber structure with a high quality factor using two circular ring resonators for biomedical sensing." *Inventions*, 6, no. 4, 78.

[16] Kiani, Sina, Pejman Rezaei, Moein Navaei, and Mohammad Sadegh Abrishamian. (2018). "Microwave sensor for detection of solid material permittivity in single/multilayer samples with high quality factor." *IEEE Sensors Journal*, 18, no. 24, 9971-9977.

[17] Nickpay, Mohammad-Reza, Mohammad Danaie, and Ali Shahzadi. (2022). "Highly sensitive THz refractive index sensor based on folded split-ring metamaterial graphene resonators." *Plasmonics*, 17, no. 1, 237-248.

[18] Kiani, Sina, Pejman Rezaei, Mahmood Karami, and Ramazan Ali Sadeghzadeh. (2019). "Band-stop filter sensor based on SIW cavity for the non-invasive measuring of blood glucose." *IET Wireless Sensor Systems*, 9, no. 1, 1-5.

[19] Karami, M., Rezaei, P., Kiani, S., and Sadeghzadeh, R. A. (2017). "Modified planar sensor for measuring dielectric constant of liquid materials." *Electronics Letters*, 53, no. 19, 1300-1302.

[20] Kiani, Sina, Pejman Rezaei, and Mina Fakhr. (2021). "Dual-frequency microwave resonant sensor to detect noninvasive glucose-level changes through the fingertip." *IEEE Transactions on Instrumentation and Measurement*, 70, 1-8.

[21] Ghods, Mohammad Mahdi, and Pejman Rezaei. (2018). "Graphene-based Fabry-Perot resonator for chemical sensing applications at mid-infrared frequencies." *IEEE Photonics Technology Letters*, 30, no. 22, 1917-1920.

[22] Kiani, Sina, Pejman Rezaei, and Moein Navaei. (2020). "Dual-sensing and dual-frequency microwave SRR sensor for liquid samples permittivity detection." *Measurement*, 160, 107805.

[23] Viktor, G. VESELAGO. (1968). "The electrodynamics of substances with simultaneously negative values of ε and μ." *Soviet Physics Uspekhi*, 10, no. 4, 509.

[24] Gong, Shaokang, Binggang Xiao, Lihua Xiao, Shengjun Tong, Sanshui Xiao, and Xiumin Wang. (2019). "Hybridization-induced dual-band tunable graphene metamaterials for sensing." *Optical Materials Express*, 9, no. 1, 35-43.

[25] Xiao, Shuyuan, Tao Wang, Yuebo Liu, Chen Xu, Xu Han, and Xicheng Yan. (2016). "Tunable light trapping and absorption enhancement with graphene ring arrays." *Physical Chemistry Chemical Physics*, 18, no. 38, 26661-26669.

[26] Liu, Tingting, Zao Yi, and Shuyuan Xiao. (2017). "Active control of near-field coupling in a terahertz metal-graphene metamaterial." *IEEE Photonics Technology Letters*, 29, no. 22, 1998-2001.

[27] Xiao, Shuyuan, Tao Wang, Tingting Liu, Xicheng Yan, Zhong Li, and Chen Xu. (2018). "Active modulation of electromagnetically induced transparency analogue in terahertz hybrid metal-graphene metamaterials." *Carbon*, 126, 271-278.

[28] Novoselov, Kostya S., Andre K. Geim, Sergei V. Morozov, De-eng Jiang, Yanshui Zhang, Sergey V. Dubonos, Irina V. Grigorieva, and Alexandr A. Firsov. (2004). "Electric field effect in atomically thin carbon films." *Science*, 306, no. 5696, 666-669.

[29] Liu, Lin, Jiajia Chen, Zigang Zhou, Zao Yi, and Xin Ye. (2018). "Tunable absorption enhancement in electric split-ring resonators-shaped graphene arrays." *Materials Research Express*, 5, no. 4, 045802.

[30] Zeng, Yu, XiFang Chen, Zao Yi, Yougen Yi, and Xibin Xu. (2018). "Fabrication of pn heterostructure ZnO/Si moth-eye structures: Antireflection, enhanced charge separation and photocatalytic properties." *Applied Surface Science*, 441, 40-48.

[31] Liu, Peter Q., Isaac J. Luxmoore, Sergey A. Mikhailov, Nadja A. Savostianova, Federico Valmorra, Jérôme Faist, and Geoffrey R. Nash. (2015). "Highly tunable hybrid metamaterials employing split-ring resonators strongly coupled to graphene surface plasmons." *Nature communications*, 6, no. 1, 1-7.

[32] Gusynin, V. P., Sharapov, S. G., and Carbotte, J. P. (2006). "Magneto-optical conductivity in graphene." *Journal of Physics: Condensed Matter*, 19, no. 2, 026222.

[33] Gao, Huajing, Fei Wang, Shifa Wang, Xiangxian Wang, Zao Yi, and Hua Yang. (2019). "Photocatalytic activity tuning in a novel Ag2S/CQDs/CuBi2O4 composite: Synthesis and photocatalytic mechanism." *Materials Research Bulletin*, 115, 140-149.

[34] Huang, Mulin, Yongzhi Cheng, Zhengze Cheng, Haoran Chen, Xuesong Mao, and Rongzhou Gong. (2018). "Based on graphene tunable dual-band terahertz metamaterial absorber with wide-angle." *Optics Communications*, 415, 194-201.

[35] Hanson, George, W. (2008). "Dyadic Green's functions and guided surface waves for a surface conductivity model of graphene." *Journal of Applied Physics*, 103, no.6, 064302.

[36] Meng, Tianhua, Dan Hu, and Qiaofen Zhu. (2018). "Design of a five-band terahertz perfect metamaterial absorber using two resonators." *Optics Communications*, 415, 151-155.

[37] Appasani, Bhargav, Pallav Prince, Rajeev Kumar Ranjan, Nisha Gupta, and Vijay Kumar Verma. (2019). "A simple multi-band metamaterial absorber with combined polarization sensitive and polarization insensitive characteristics for terahertz applications." *Plasmonics*, 14, no. 3, 737-742.

[38] Bakshi, Saikat Chandra, Debasis Mitra, and Laxmikant Minz. (2018). "A compact design of multiband terahertz metamaterial absorber with frequency and polarization tunability." *Plasmonics*, 13, no. 6, 1843-1852.

[39] Zamzam, Pouria, and Pejman Rezaei. (2021). "A terahertz dual-band metamaterial perfect absorber based on metal-dielectric-metal multi-layer columns." *Optical and Quantum Electronics*, 53, no. 2, 1-9.

[40] Chen, Ming, Wei Sun, Jianjin Cai, Linzi Chang, and Xiaofei Xiao. (2017). "Frequency-tunable terahertz absorbers based on graphene metasurface." *Optics Communications*, 382, 144-150.

[41] Qi, Yunping, Yu Zhang, Chuqin Liu, Ting Zhang, Baohe Zhang, Liyuan Wang, Xiangyu Deng, Yulong Bai, and Xiangxian Wang. (2020). "A tunable terahertz metamaterial absorber composed of elliptical ring graphene arrays with refractive index sensing application." *Results in Physics*, 16, 103012.

[42] Zamzam, Pouria, Pejman Rezaei, and Seyed Amin Khatami. (2021). "Quad-band polarization-insensitive metamaterial perfect absorber based on bilayer graphene metasurface." *Physica E: Low-dimensional Systems and Nanostructures*, 128, 114621.

[43] Li, Wangyang, and Yongzhi Cheng. (2020). "Dual-band tunable terahertz perfect metamaterial absorber based on strontium titanate (STO) resonator structure." *Optics Communications*, 462, 125265.

[44] Huang, Mulin, Yongzhi Cheng, Zhengze Cheng, Haoran Chen, Xuesong Mao, and Rongzhou Gong. (2018). "Based on graphene tunable dual-band terahertz metamaterial absorber with wide-angle." *Optics Communications*, 415, 194-201.

[45] Wang, Ben-Xin, Zhai, X., Wang, G. Z., Huang, W. Q., and Wang, L. L. (2014). "Design of a four-band and polarization-insensitive terahertz metamaterial absorber." *IEEE Photonics Journal*, 7, no. 1, 1-8.

[46] Arezoomand, Afsaneh Saee, Ferdows B. Zarrabi, Samaneh Heydari, and Navid P. Gandji. (2015). "Independent polarization and multi-band THz absorber base on Jerusalem cross." *Optics Communications*, 352, 121-126.

[47] Qin, Feng, Zeqiang Chen, Xifang Chen, Zao Yi, Weitang Yao, Tao Duan, Pinghui Wu, Hua Yang, Gongfa Li, and Yougen Yi. (2020). "A tunable triple-band near-infrared metamaterial absorber based on Au nano-cuboids array." *Nanomaterials*, 10, no. 2, 207.

[48] Wang, Ben-Xin, and Gui-Zhen Wang. (2016). "Quad-band terahertz absorber based on a simple design of metamaterial resonator." *IEEE Photonics Journal*, 8, no. 6, 1-8.

[49] Huang, Xiutao, Conghui Lu, Cancan Rong, Zhaoyang Hu, and Minghai Liu. (2018). "Multiband ultrathin polarization-insensitive terahertz perfect absorbers with complementary metamaterial and resonator based on high-order electric and magnetic resonances." *IEEE Photonics Journal*, 10, no. 6, 1-11.

[50] Shan, Yan, Lin Chen, Cheng Shi, Zhaoxiang Cheng, Xiaofei Zang, Boqing Xu, and Yiming Zhu. (2015). "Ultrathin flexible dual band terahertz absorber." *Optics Communications*, 350, 63-70.

[51] Yi, Zao, Hang Lin, Gao Niu, Xifang Chen, Zigang Zhou, Xin Ye, Tao Duan, Yong Yi, Yongjian Tang, and Yougen Yi. (2019). "Graphene-based tunable triple-band plasmonic perfect metamaterial absorber with good angle-polarization-tolerance." *Results in Physics*, 13, 102149.

[52] Yuan, Su, Rongcao Yang, Jinping Tian, and Wenmei Zhang. (2020). "A photoexcited switchable tristate terahertz metamaterial absorber." *International Journal of RF and Microwave Computer-Aided Engineering*, 30, no. 1, e22014.

[53] Fan, Yuenong. (2019). "Dual-band perfect light absorber in visible region based on cylinder silicon resonator." *Optik*, 179, 1084-1090.

[54] Wang, Ben-Xin, Yuanhao He, Pengcheng Lou, Wei-Qing Huang, and Fuwei Pi. (2020). "Penta-band terahertz light absorber using five localized resonance responses of three patterned resonators." *Results in Physics*, 16, 102930.

[55] Xiang, Yulin, Lingling Wang, Qi Lin, Shengxuan Xia, Meng Qin, and Xiang Zhai. (2019). "Tunable dual-band perfect absorber based on L-shaped graphene resonator." *IEEE Photonics Technology Letters*, 31, no. 6, 483-486.

[56] Nejat, Mohamad, and Najmeh Nozhat. (2019). "Ultrasensitive THz refractive index sensor based on a controllable perfect MTM absorber." *IEEE sensors journal*, 19, no. 22, 10490-10497.

Chapter 9

Highly Sensitive and Polarization-Insensitive Terahertz Metamaterial-Based Sensor Using Four Circular Ring Resonators

Pritha Mukheerjee[1], Snehashis Basu[1], Purba Dutta[1], Sagnik Banerjee[1], Amitkumar V. Jha[1], Bhargav Appasani[1] and Yadgar Abdulkarim[2,*]

[1]Kalinga Institute of Industrial Technology, Bhubaneshwar, India
[2]Medical Physics Department, College of Medicals & Applied Science, Charmo University, Sulaimania, Iraq

Abstract

In this chapter, following the enormous applicability and literature surrounding metamaterial absorbers, a unique design in the terahertz regime is proposed. Using gold as the reflective metal, gallium arsenide as the photoconductive substrate and the novel design assures a high and distinct absorption rate of 99%. The absorber can also sense the changes in the refractive index of the medium around it. The absorption spectrum shifts with the changes in refractive index, enabling it to sense the medium. The proposed sensor attains a high sensitivity of 2.558 THz/RIU over six consecutive values of refractive indices ranging from 1 to 1.05 with a step size of 0.01. The range for the refractive index has been chosen to provide solid scope to detect multiple types of gases in the surrounding medium and thereby act as a potential gas sensor. A large quality factor and figure of merit of 283.66 and 255.8 RIU-1, respectively, ensure the sensor's impressive performance.

---

* Corresponding Author's Email: Yadgar.kharkov@gmail.com.

In: Metamaterials for Microwave and Terahertz Applications
Editors: B. Appasani, O. Prakash Acharya, A. Vidyakant Jha et al.
ISBN: 979-8-88697-330-3

Keywords: metamaterial absorber, THz, refractive index sensor, gas sensor, sensitivity

1. Introduction

The laws of optics and electromagnetics can be tuned and manipulated to attain stellar results by using periodic sub-wavelength structures called metamaterials [1, 2]. Such tuneability has led to a wide range of applications in biosensing, imaging, optoelectronics, and antenna engineering [3-6]. Metamaterial absorbers (MA) work on the principle that the absorbed incident wave converts into thermal energy leading to wave attenuation [7, 8]. By detecting this absorption, the object under study can be sensed. The terahertz (THz) frequency regime offers a large bandwidth. Its low scattering and shorter wavelengths are useful for sensing and imaging applications. In addition, because of its non-ionizing nature, the waves do not interfere with the absorber. Photoconductive dielectrics like gallium arsenide (GaAs) [9], with their large bandgap, are highly suitable for THz applications. GaAs is also profoundly used in constructing the MAs and reflective metals like gold and silver to act as resonators.

The use of refractive index (RI) sensors has wide applications in several domains like cancer cell detection, detection of analyte concentration, and gas sensing [10-13]. An innovative MTM-based sensor has been reported in [14] to detect the presence of transition metals in S and C frequency bands. An RI sensor based on metasurfaces [15] yielded a very narrow bandwidth of 1.74 nm. The high sensitivity obtained for this sensor model has been owed to the carved-out cross in the silver substrate. While most MTM-based models work under the principle of surface plasmon resonance [16], the MA designed in [17] for RI sensing uses Fabry-Pérot resonance to attain an absorbance rate of 99% in the near-infrared frequency region. A growing area of focus in high-frequency applications is the use of graphene. Its high stability yields high performance in a wide range of applications. A gold-coated graphene substrate has been proposed in [16] for RI sensing in the THz range. Du et al. [18] proposed a graphene-based MA that yielded a high sensitivity of 1.84 THz/RIU. With an alarming rise in air pollutants and the global warming crisis, the need to detect harmful greenhouse gases is quintessential [19, 20].

The RI sensor proposed in this paper paves a way to detect such gases by sensing their corresponding refractive indices. The high sensitivity and FoM obtained can have wide applications in several domains. Table 1 depicts the performance of the present study through comparison with previous related works.

The proposed sensor uses gold as the metal resonators and GaAs as the dielectric substrate in the frequency range of 0.1-5 THz. The sensor's design is shown in the next section. After that, in Section III, we present its measures for the performance shown with the simulation results.

Table 1. Comparison of Sensor Performance with previous works

Ref. No.	Q-factor	Sensitivity (GHz/RIU)	FoM
[21]	78.90	0.3537	11.0531
[22]	13.76	851	2.927
[23]	32.167	187	6.015
[24]	44	1500	25
[25]	29.5, 66.8, 59.4	540, 700, 1500	6.75, 17.5, 30
[26]	231	186	187
This paper	283.66	2558	255.8

2. Design and Simulation

Figure 1 shows the structural design of the sensor proposed in this paper. The sensor consists of a ground plane and resonating structures at the top, both of which are gold (conductivity, $\sigma = 4.1 \times 10^7$ S/m). The dielectric substrate layer between the top and bottom structures is gallium arsenide (GaAs, loss tangent $\tan\delta = 6 \times 10^{-3}$ and relative permittivity $\varepsilon_r = 12.94$). The top surface of the sensor has a thickness, $t = 0.2$ µm, placed on a unit cell of dimensions, $u = 100$ µm, and height, $h = 5$ µm. It consists of four circular ring resonators (CRRs) with a radius $r = 12$ µm. A low height of the substrate ensures mechanical flexibility in the THz region. Periodic boundary conditions are applied along the x and y axes, while incident transverse electromagnetic (TEM) waves propagate along the z-axis.

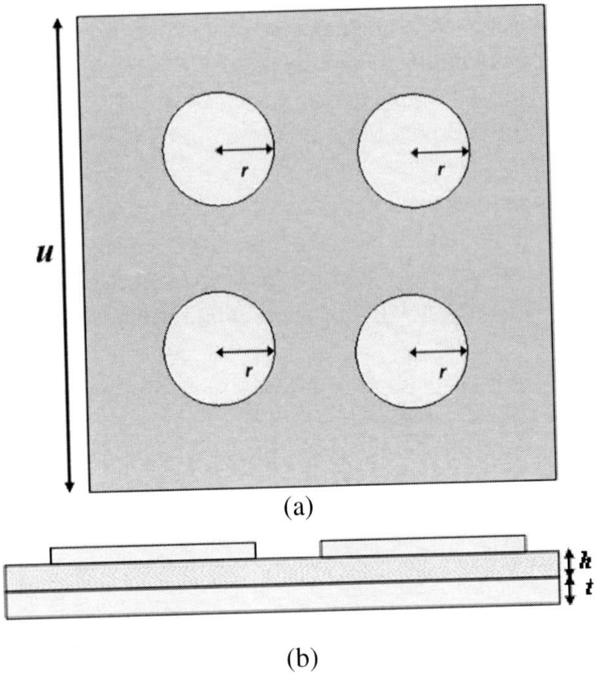

Figure 1. Structural design of the proposed sensor in its (a) front view and (b) bottom view.

3. Results and Analysis

The design and all corresponding simulations for the sensor have been performed using CST Studio Professional v. 2019. The ground plane prevents transmission through the model, that is, the transmission coefficient, T = 0. Absorption, A is given by the expression,

$$A = 1 - R - T = 1 - |S_{11}|^2 - |S_{21}|^2 \tag{1}$$

where, S_{11} and S_{21} denote the S-parameters for reflection and transmission, respectively. Hence, the reflection coefficient R should be minimized as much as possible to obtain maximum absorption. Our model has an absorption rate of 99% with a (FWHM) of 0.01 THz at 2.836 THz, as shown in Figure 2. In sensing and imaging applications, the effects of polarization lead to a low signal-to-noise ratio (SNR), distorting the results [27, 28]. The symmetry of the proposed structure makes the sensor polarization independent. This is shown in Figure 3, where the absorption characteristics have negligible change with change in angle of polarization, ϕ, thereby preventing any hindrance.

The working mechanism of the proposed gas sensor can be analyzed through the surface current distribution in Figure 4. The surface current is denser along the edges of the CRRs and over the distance between the two pairs of horizontally placed resonators. The similarity of the distribution patterns of each resonator also proves that they provide an equal and exclusive contribution to the resonance mechanism.

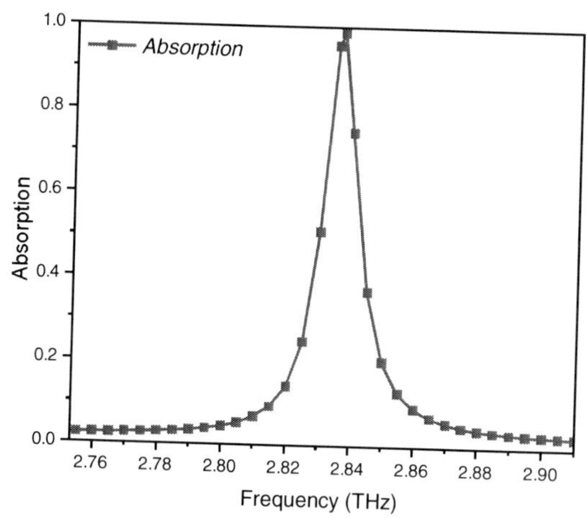

Figure 2. Absorption spectrum for the gas sensor.

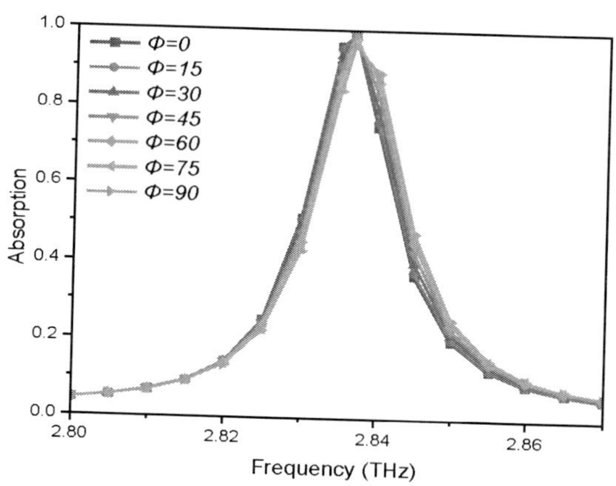

Figure 3. Polarisation independence of the sensor.

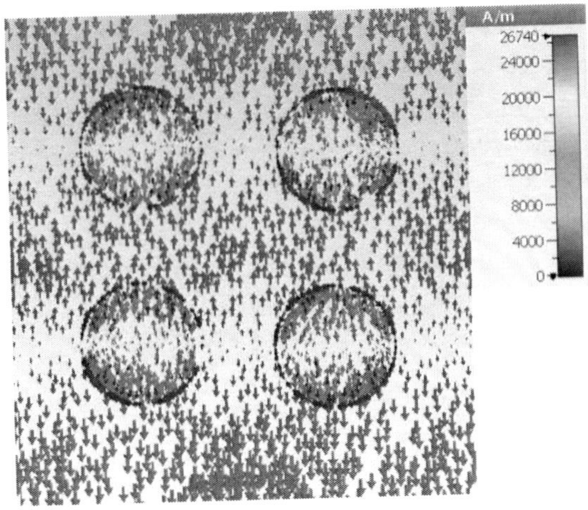

Figure 4. Surface Current Distribution.

To justify our choice of values for the parameters, parametric sweep for the unit cell dimension, *u*, the height of substrate, h, and the radius of the CRRs, *r*. As is shown in Figure 5, the parametric sweep has been performed for three consecutive integer values for each parameter. In selecting the best results for a sensor model, the Q-factor, given by $Q = f_0/FWHM$, and sensitivity, are the two most defining measures to be considered. While a higher absorption rate is attained for u = 101 μm than for our chosen u = 100 μm, FWHM for the latter is smaller, thus producing a higher quality factor of 283.66 (Figure 5(a)). It can be inferred that there is a phase-magnitude mismatch for *u* at 101 μm and 99 μm, which leads to destructive interference at resonance. This leads to the poorer performance of the model. At h = 5 μm, impedance matching occurs, i.e., the impedance of the model becomes equal to that of free space (≈377Ω). As a result, an absorption spectrum more defined than for any of the other two values of *h* has been attained, as is shown in Figure 5(b). Resonance also depends on the distance between the resonators and can therefore be observed by varying the radius value. As the radius increases, the area occupied by the CRRs also increases, resulting in a red shift of the resonance frequency. We obtain a distinct peak only at r = 5 μm (Figure 5(c)), thus justifying the results obtained in Figure 2. The various absorption parameters for the performed parametric sweeps have been listed in Table 2 for better depiction.

Metamaterial-based RI sensors can be effectively used in fields like biosensing because of their insensitivity to external factors. In addition to polarization-insensitivity, the measurement of the RI for the present model remains unscathed by external changes like those of temperature and pressure. This is because, along with a wide bandgap, GaAs substrate has high electron mobility, making it suitable for high-temperature applications. Gold, being one of the densest metals, encounters negligible change concerning temperature changes. The combined outcome results in a high-performing sensor.

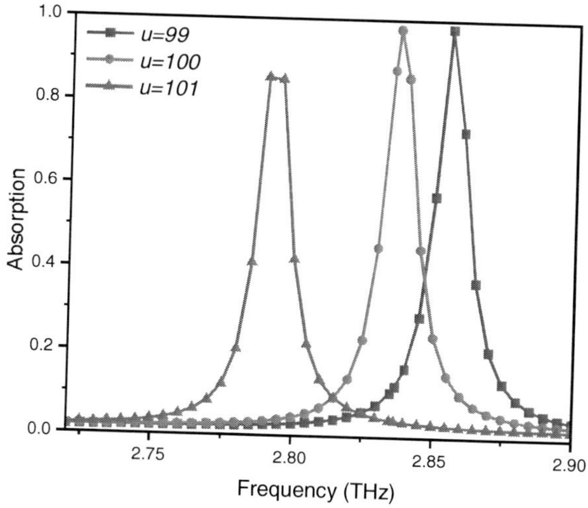

(a)

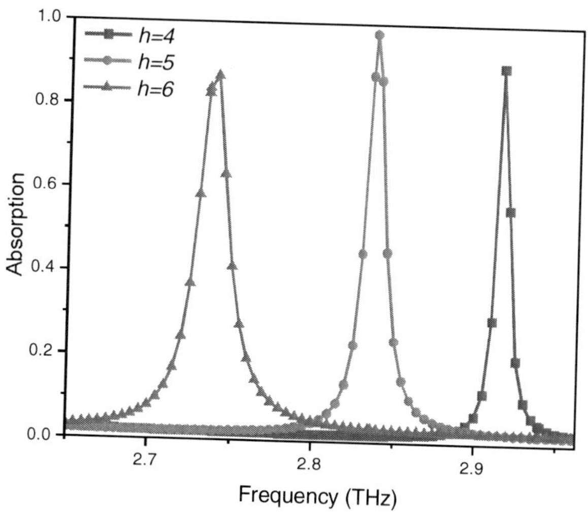

(b)

Figure 5. (Continued)

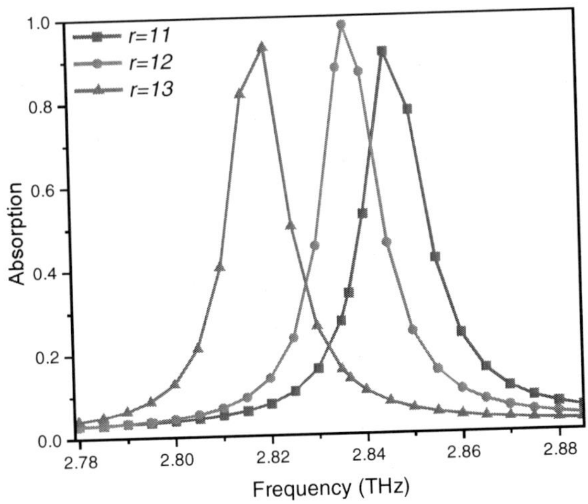

(c)

Figure 5. Justification of parameters by the parametric sweep of (a) unit cell dimension, u, (b) height, h, and (c) radius, r.

Table 2. Comparison of parameter values after Parametric Sweep

	(all units in μm)	**Resonant frequency, f_0 (THz)**	**Absorption rate, *A***	**FWHM (THz)**	**Q-factor**
Unit cell dimension, *u*	99	2.855	0.99	0.03	95.167
	100	2.836	0.99	0.01	283.66
	101	2.785	0.84	0.02	139.25
Height of substrate, *h*	4	2.92	0.86	0.01	292
	5	2.836	0.99	0.01	283.66
	6	2.75	0.84	0.02	137.5
Radius, *r*	11	2.845	0.90	0.04	71.125
	12	2.836	0.99	0.01	283.66
	13	2.82	0.91	0.04	70.5

The model's performance for sensing applications can be evaluated by observing the absorption spectrum for a range of refractive indices, as demonstrated in Figure 6. Here we have considered RI from 1-1.05 with a step width of 0.01. A significant resonant frequency shift occurs for such a small step width. This change of resonant frequency with respect to the change in RI is termed the sensitivity (S) and marks the overall performance of the model (Figure 7). It is given by the magnitude of the slope of the negatively sloped linear plot between the RI and the resonant frequencies, the equation of which is given by,

$$f = -2.558n + 5.394 \tag{2}$$

The third important metric for performance is the corresponding figure of merit (FoM), given by $S/FWHM$. The proposed model achieved a high sensitivity of 2.558 THz/RIU and a FoM of 255.8/RIU.

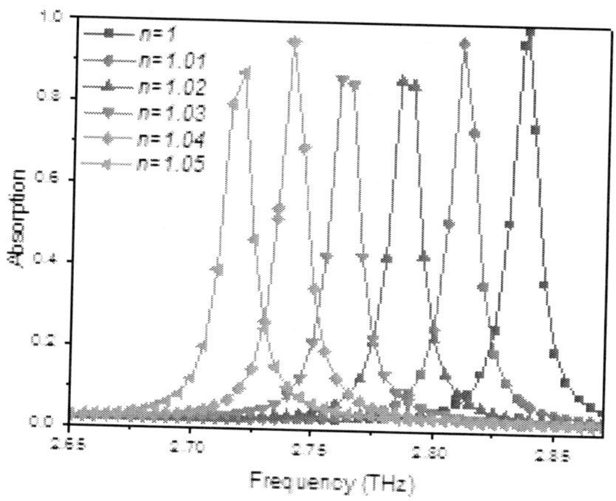

Figure 6. Variation of absorption characteristics for varying values of refractive index.

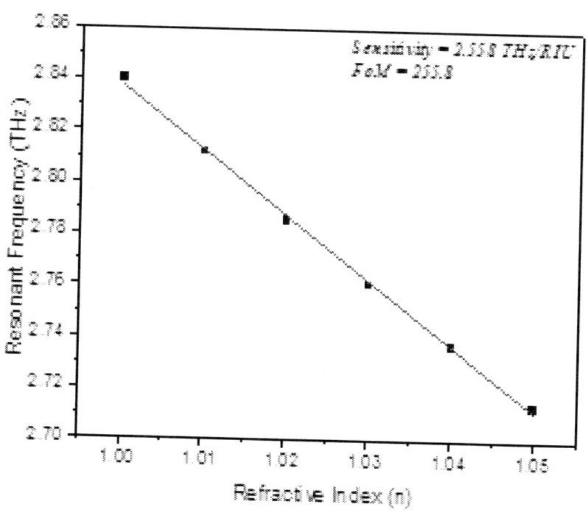

Figure 7. Sensitivity curve and figure of merit (FoM).

Table 3. Refractive Indices of Gases

Gases	**Refractive Index (*n*)**
Carbon monoxide	1.000338
Carbon dioxide	1.000449
Sulphur dioxide	1.000686
Chloroform	1.001450
Methane	1.000444

Most gases have refractive indices far below 1.05. Table 3 lists some common gases present in the environment and their equivalent refractive indices. Using refractive index, several gases present in the surrounding medium of the sensor can thus be detected. The high sensitivity and FoM obtained for the selected range of RI make the sensor highly fruitful in gas sensing applications.

Conclusion

The unique design of a metamaterial absorber in the THz region is presented in this paper. High sensitivity and FoM of 2.558 THz/RIU and 255.8/RIU, respectively, are obtained over a refractive index range from 1.00-1.05 with 0.01 as the step width. The efficiency of the sensor is validated by a high Q-factor value of 283.66. The sensor is lightweight and polarization-independent. The photoconductive dielectric and highly reflective circular resonators in the THz range guarantee the refractive index measurements without any distortion. The high performative measures obtained in the RI range ensures scope for RI sensing in feasible gas sensing applications.

References

[1] Veselago, V. G. "The Eelctromagnetics of Substances with Simultaneously Negative Values of and μ." *Soviet Physics Uspekhi.* 10, no. 4 (1968): 1.

[2] Smith, D. R., Willie Padilla, D. C. Vier, S. C. Nemat-Nasser, and S. Schultz. "Negative permeability from split ring resonator arrays." In *Conference Digest. 2000 Conference on Lasers and Electro-Optics Europe (Cat. No. 00TH8505)*, pp. 1-pp. IEEE, 2000.

[3] Landy, N. I., C. M. Bingham, T. Tyler, N. Jokerst, D. R. Smith, and W. J. Padilla. "Design, theory, and measurement of a polarization-insensitive absorber for terahertz imaging." *Physical Review B* 79, no. 12 (2009): 125104.

[4] Sabah, Cumali, Furkan Dincer, Muharrem Karaaslan, Emin Unal, Oguzhan Akgol, and Ekrem Demirel. "Perfect metamaterial absorber with polarization and incident angle independencies based on ring and cross-wire resonators for shielding and a sensor application." *Optics Communications* 322 (2014): 137-142.

[5] Rashed, Ahmed Nabih Zaki, Abd El-Naser A. Mohammed, Walid Fawzy Zaky, I. S. Amiri, and P. Yupapin. "The switching of optoelectronics to full optical computing operations based on nonlinear metamaterials." *Results in Physics* 13 (2019): 102152.

[6] Bağmancı, Mehmet, Muharrem Karaaslan, Emin Ünal, Oguzhan Akgol, and Cumali Sabah. "Extremely-broad band metamaterial absorber for solar energy harvesting based on star shaped resonator." *Optical and Quantum Electronics* 49, no. 7 (2017): 1-14.

[7] Yan, Zhendong, Chaojun Tang, Guohua Wu, Yumei Tang, Ping Gu, Jing Chen, Zhengqi Liu, and Zhong Huang. "Perfect absorption and refractive-index sensing by metasurfaces composed of cross-shaped hole arrays in metal substrate." *Nanomaterials* 11, no. 1 (2020): 63.

[8] Stanley, Ross. "Plasmonics in the mid-infrared." *Nature Photonics* 6, no. 7 (2012): 409-411.

[9] Tani, Masahiko, Shuji Matsuura, Kiyomi Sakai, and Shin-ichi Nakashima. "Emission characteristics of photoconductive antennas based on low-temperature-grown GaAs and semi-insulating GaAs." *Applied optics* 36, no. 30 (1997): 7853-7859.

[10] La Spada, Luigi, Filiberto Bilotti, and Lucio Vegni. "Metamaterial biosensor for cancer detection." In *SENSORS, 2011 IEEE*, pp. 627-630. IEEE, 2011.

[11] Azab, Mohammad Y., Mohamed Farhat O. Hameed, Abed M. Nasr, and S. S. A. Obayya. "Highly sensitive metamaterial biosensor for cancer early detection." *IEEE Sensors Journal* 21, no. 6 (2021): 7748-7755.

[12] Dixit, A., S. Tiwari, U. Ramani, and P. C. Pandey. "Refractive index sensor based on evanescent field effects in hollow core PCF for detection of analytes over extended E+S+C+L+U communication bands." *Optics & Laser Technology* 121 (2020): 105779.

[13] Monfared, Yashar E., and Montasir Qasymeh. "Graphene-assisted infrared plasmonic metamaterial absorber for gas detection." *Results in Physics* 23 (2021): 103986.

[14] Ma, Lihua, Zijian Cui, Dongying Zhu, Lisha Yue, Lei Hou, and Yue Wang. "Metamaterials sensor based on multiband terahertz absorber." In *2019 44th International Conference on Infrared, Millimeter, and Terahertz Waves (IRMMW-THz)*, pp. 1-2. IEEE, 2019.

[15] Yan, Zhendong, Chaojun Tang, Guohua Wu, Yumei Tang, Ping Gu, Jing Chen, Zhengqi Liu, and Zhong Huang. "Perfect absorption and refractive-index sensing by metasurfaces composed of cross-shaped hole arrays in metal substrate." *Nanomaterials* 11, no. 1 (2020): 63.

[16] Xiang, Yuanjiang, Jiaqi Zhu, Leiming Wu, Qi You, Banxian Ruan, and Xiaoyu Dai. "Highly sensitive terahertz gas sensor based on surface plasmon resonance with graphene." *IEEE Photonics Journal* 10, no. 1 (2017): 1-7.

[17] Kumar, Raghwendra, and Pankaj Kumar. "Fabry-Pérot cavity resonance based metamaterial absorber for refractive index sensor at infrared frequencies." *Optics Communications* 514 (2022): 128142.

[18] Du, Xuemei, Fengping Yan, Wei Wang, Luna Zhang, Zhuoya Bai, Hong Zhou, and Yafei Hou. "Thermally-stable graphene metamaterial absorber with excellent tunability for high-performance refractive index sensing in the terahertz band." *Optics & Laser Technology* 144 (2021): 107409.

[19] Akimoto, Hajime. "Global air quality and pollution." *Science* 302, no. 5651 (2003): 1716-1719.

[20] Kumar, Prashant, Lidia Morawska, Claudio Martani, George Biskos, Marina Neophytou, Silvana Di Sabatino, Margaret Bell, Leslie Norford, and Rex Britter. "The rise of low-cost sensing for managing air pollution in cities." *Environment international* 75 (2015): 199-205.

[21] Zhang, Wei, Jian-Ying Li, and Jian Xie. "High sensitivity refractive index sensor based on metamaterial absorber." *Progress In Electromagnetics Research M* 71 (2018): 107-115.

[22] Nickpay, Mohammad-Reza, Mohammad Danaie, and Ali Shahzadi. "Highly sensitive THz refractive index sensor based on folded split-ring metamaterial graphene resonators." *Plasmonics* 17, no. 1 (2022): 237-248.

[23] Banerjee, Sagnik, Uddipan Nath, Amitkumar V. Jha, Sasmita Pahadsingh, B. Appasani, Nicu Bizon, and Avireni Srinivasulu. "A terahertz metamaterial absorber based refractive index sensor with high quality factor." In *2021 13th International Conference on Electronics, Computers and Artificial Intelligence (ECAI)*, pp. 1-4. IEEE, 2021.

[24] Banerjee, Sagnik, Uddipan Nath, Purba Dutta, Amitkumar Vidyakant Jha, Bhargav Appasani, and Nicu Bizon. "A theoretical terahertz metamaterial absorber structure with a high quality factor using two circular ring resonators for biomedical sensing." *Inventions* 6, no. 4 (2021): 78.

[25] Banerjee, Sagnik, Purba Dutta, Amitkumar V. Jha, Prabhat R. Tripathi, Avireni Srinivasulu, Bhargav Appasani, and Cristian Ravariu. "A Triple Band Highly Sensitive Refractive Index Sensor Using Terahertz Metamaterial Perfect Absorber." *Progress In Electromagnetics Research M* 107 (2022): 13-24.

[26] Wen, Yao, Tao Xu, and Yu-Sheng Lin. "Design of electrostatically tunable terahertz metamaterial with polarization-dependent sensing characteristic." *Results in Physics* 29 (2021): 104798.

[27] Li, Songquan, Laixu Gao, Changwei Zou, Wei Xie, Yong Wei, Canxin Tian, Zesong Wang, Feng Liang, Yanxiong Xiang, and Qian Yang. "A polarization-independent fiber-optic SPR sensor." *Sensors* 18, no. 10 (2018): 3204.

[28] Abdelsalam, Mostafa, Ahmed M. Mahmoud, and Mohamed A. Swillam. "Polarization independent dielectric metasurface for infrared beam steering applications." *Scientific reports* 9, no. 1 (2019): 1-7.

About the Editors

Bhargav Appasani, PhD (Engg.), MIEEE
Assistant Professor (II) and Research Faculty,
School of Electronics Engineering,
Kalinga Institute of Industrial Technology,
Bhubaneswar, India
Email: bhargav.appasanifet@kiit.ac.in

Om Prakash Acharya
School of Electronics Engineering, Kalinga Institute of Industrial Technology, Bhubaneswar, India
Email: omprakash.acharyafet@kiit.ac.in

Amitkumar Vidyakant Jha
School of Electronics Engineering, Kalinga Institute of Industrial Technology, Bhubaneswar, India
Email: amit.jhafet@kiit.ac.in

Nisha Gupta
Department of Electronics and Communication Engineering, Birla Institute of Technology, Mesra, India
Email: ngupta@bitmesra.ac.in

Index

P

R

S

T

U

V

W